Essentials of Neuropsychology

This comprehensive textbook offers a holistic integration of both the research and clinical aspects of neuropsychology. Combining Eastern and Western perspectives, it explores latest developments, current scope, and significant challenges in the field to provide a detailed understanding of brain and behavior from research and intervention methods to rehabilitation and applications. Each chapter in the book includes an introduction to the topic, an overview of the latest research in the field, and a discussion of the future directions that research can take.

The book is structured into three parts, each addressing specific aspects of the field. Part 1 introduces the readers to the fundamental principles of neuropsychology, including the available methods of assessment, brain anatomy, and its connection with human psychology. It provides an in-depth look at neuropsychological and electrophysiological methods and their applications in clinical practice. Part 2 focuses on the brain and cognition, examining the complex mechanisms that underlie cognitive behavior. The chapters include neuropsychology of various executive functions, memory, and social cognition. Part 3 delves into clinical disorders and their biological basis. This section explores the disorders that have a direct relationship between brain functioning and behavior, offering valuable insights into their diagnosis, treatment, and management.

It is an essential resource for both students in clinical neuropsychology and professionals seeking to expand their knowledge and stay abreast of the latest developments.

K. Jayasankara Reddy is a Professor in the Department of Psychology & Co-Ordinator for the Centre for Research at Christ University, Bangalore, India, as well as an active teacher, researcher, and practitioner in the field of Health, Cognitive Neuropsychology, and Neuroscience.

Essentials of Neuropsychology
Integrating Eastern and Western Perspectives

K. Jayasankara Reddy

Routledge
Taylor & Francis Group

LONDON AND NEW YORK

Designed cover image: © Getty Images\AlexSava

First published 2024
by Routledge
4 Park Square, Milton Park, Abingdon, Oxon OX14 4RN

and by Routledge
605 Third Avenue, New York, NY 10158

Routledge is an imprint of the Taylor & Francis Group, an informa business

© 2024 K. Jayasankara Reddy

British Library Cataloguing-in-Publication Data
A catalogue record for this book is available from the British Library

Library of Congress Cataloging-in-Publication Data
Names: Reddy, K. Jayasankara, author.
Title: Essentials of neuropsychology : integrating eastern and western perspectives / Dr. K. Jayasankara Reddy.
Description: 1 Edition. | New York, NY : Routledge, 2023. | Includes bibliographical references and index.
Identifiers: LCCN 2023037817 | ISBN 9781032639789 (paperback) | ISBN 9781032640822 (hardback) | ISBN 9781032640839 (ebook)
Subjects: LCSH: Neuropsychology—Textbooks. | Cognitive neuroscience—Textbooks. | Cognition—Textbooks. | Nervous system—Diseases—Textbooks.
Classification: LCC QP360 .R423 2023 | DDC 612.8/233—dc23/eng/20231026
LC record available at https://lccn.loc.gov/2023037817

ISBN: 978-1-032-64082-2 (hbk)
ISBN: 978-1-032-63978-9 (pbk)
ISBN: 978-1-032-64083-9 (ebk)

DOI: 10.4324/9781032640839

Typeset in Galliard
by codeMantra

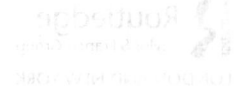

Contents

Acknowledgments

I am expressing my sincere gratitude to the Center for Research Projects at Christ (Deemed to be University), Bangalore for the generous funding and required assistance during the writing process. I am grateful to the Vice Chancellor for his invaluable support and encouragement in completing this book. I extend my gratitude to the book's editors, whose valuable insights and feedback have been instrumental in developing the book into what it is today. I sincerely thank the reviewers; their constructive criticism and suggestion helped refine the book. Finally, a note of gratitude to the research assistants Adithya Ramesh and Shuchita Gupta; this book would not have been possible without their research, hard work, and dedication. Once again, I would like to thank everyone who has contributed to this book, whether directly or indirectly. Your support and encouragement have been instrumental throughout the writing process.

Dr. K. Jayasankara Reddy PhD,
Professor of Neuropsychology & Neuroscience
Department of Psychology,
Christ University, Bangalore 560 029, India

Part 1

1 Neuropsychology and Its Beginnings

Introduction

The human body consists of several vital organs. It is imperative to acknowledge that each organ is essential for our survival, and none are expendable. Everything exists in a delicate balance that can only be truly explained as a result of billions of years of evolution. The human species is not unlike any other species that inhabit this rock that revolves around a giant ball of flame. The very fact that we share about 96% of our genes with chimpanzees means that we are not too far along the evolutionary scale. In the timeline, we are merely a blip. Our sense of deluded superiority comes from the fact that we are the owners of large brains. In comparison to our closest relatives, the chimpanzees, we have a much larger brain. Notably, we have a well-developed cerebral cortex which has also led to establishing higher-level cognitive functioning in humans (Mora-Bermúdez et al., 2016).

As a result, this has led to an increase in research on the functions of the brain and what it can potentially do. Many brain myths have now been debunked as techniques and methods to study the brain keep developing. A popular one has always been the fact that we only use about 10% of our brains at any point in time. This has been the result of the equipotentiality theory and also pop-culture movies like *Lucy* (2014) starring popular Hollywood Actress Scarlett Johansson.

The discipline of neuropsychology is relatively new in the field of psychology. Its development can be traced to the third dynasty of ancient Egypt (Finger, 2000). Earlier, the brain was considered futile and was discarded during anatomical procedures like autopsies. With the development gained due to the understanding of human anatomy and physiology, different theories were developed regarding the functioning of the body. The brain was initially not considered the hub of bodily functions and was often viewed from a religious perspective. All kinds of abnormalities were attributed to the presence of spirits. However, it has taken several years for us to gain an understanding of the brain and its relation to behavior. Neuropsychology emerged as an independent area of study in the 1960s. According to Meier (1974), neuropsychology is "the scientific study of brain-behaviour relationship." Berlucchi (2009)

DOI: 10.4324/9781032640839-2

defined neuropsychology as the discipline which aims to understand the relationship between the brain, on the one hand, and the "mind" and behavioral control, on the other. Neuropsychology is a multidisciplinary field of study combining information from fields such as psychology, neurology, biology, physiology, and pharmacology (A. Benton, 1988; Humaida, 2019).

Sub-Disciplines of Neuropsychology

In the nascent years, neuropsychology researchers primarily used primates, patients with brain lesions, and degenerative disorders to observe and study behavioral dysfunction. Using the methods available then, researchers mostly examined single cases to gain insight into the interplay between the brain and behavior. In the last several decades, with the advancement of all-rounded technologies, researchers have been able to delve deeper into investigating the intricate relationship between mind, brain, and behavior. With such progress, the focus of the field has also broadened. Research in neuropsychology has now diversified into sub-disciplines, each emphasizing a specific aspect of the field. Clinical neuropsychology, cognitive neuropsychology, experimental neuropsychology, developmental neuropsychology, and behavioral neuropsychology are some of the major sub-disciplines within neuropsychology (Stebbins, 2007). Clinical neuropsychology is the branch that deals with the study of the etiology, assessment, and management of neurological diseases and disorders. It primarily deals with patients with brain pathology. For example, if a person suffers a traumatic brain injury (TBI), then the assessment of any functional deficits as a result of physiological damage would fall under this branch. Experimental neuropsychology is the study of human behavior as a result of brain activity and tends to depend heavily on laboratory experiments. It seeks to understand the structure and functioning of the nervous system in relation to human psychology. Developmental psychology probes into the development of the brain across the life span and how the mental process and behavior change as a result of biological changes. Researchers in this field are also working on developing interventions and rehabilitation for the well-being of individuals at different life stages. Cognitive neuropsychology addresses the brain parts and their processes underpinning cognition. This area encompasses a broad range of topics, including memory, perception, and attention. Behavioral neuropsychology is a fairly newly emerging sub-area within neuropsychology as it blends knowledge from the field of neuropsychology and behavioral therapy. Rehabilitation using the knowledge from neuropsychology is the essence of behavioral neuropsychology (Goldstein, 1990). After a brief understanding of the field, it is pertinent to gain rudimentary insight into the root and progression of the field.

Historical Background

Ancient Egypt had an inclination toward the scientific approach to abnormalities and their treatment. But they considered the heart as the seat of the soul and not the brain (Carus, 1905). Aristotle reinforced this when he attributed the heart to the control of mental processes and viewed the brain as a mechanism for cooling the heat generated by the heart (A. L. Benton & Sivan, 2007). His conclusions were based on the empirical study of animals. His beliefs regarding the working of the heart and brain remain evident even in the modern language of neuropsychology; as the saying goes that "We follow our hearts and learn by the heart" (A. L. Benton & Sivan, 2007). The transposition of this belief was initiated by the efforts of Hippocrates, who viewed the brain as the seat of the soul. He outlined the relationship between the brain and behavior by stating that "The brain exercises the greatest power in the man" (Finger, 2000). His theory initiated more scientific discoveries of the brain that is responsible for our behavior. He proposed the concept of "Mind," which was considered to function separately from the brain.

In the mid-17th century, Thomas Willis took a physiological approach to the brain and behavior. He coined the terms "hemisphere" and "lobes" and was one of the pioneers who used the terms "neurology" and "psychology" (Finger, 1994). He emphasized specialized structures of the brain and proposed a theory that higher structures account for complex functions and lower structures account for functions of automatic responses. His theories led the way for other pioneers to further work on disorders and dysfunctions in the brain. Further, Franz Joseph Gall theorized that personality was directly related to the structures within the brain. His efforts led to the invention of phrenology, in which the shape and structure of the brain were used to determine the intelligence and personality of individuals (A. Benton, 2000). His contribution led to various controversies regarding the limited capacity for higher cognitive processes. Still, his work is considered pivotal for laying the foundation in neuropsychology that flourished over the next decades.

Toward the late 19th century, Jean-Baptiste Bouillaud extended upon the idea of Gall and worked on the idea of distinct cortical regions having independent functioning. His work on speech contributed to the understanding of the localization of functions. However, Paul Broca, referred to as the "Father of Neuropsychology," made advances in the area of localized functions. His work deviates from the ideas of phrenology and provides an in-depth view of the scientific and psychological perspective of the brain (Cubelli & De Bastiani, 2011). Further, Lashley's contribution to the concept of equipotentiality provided evidence for plasticity in the brain. This contribution has drastically contributed to the field over the next few decades.

Neuropsychology has evolved from a localization model of cerebral functioning where deficits were linked to specific behavioral domains to a localized area of a brain lesion. This is practiced by neuropsychologists even at present while localizing any function or anomaly of the brain (Lassonde et al., 2006).

It was with Paul Broca that functional specialization was found. A patient was known as *Tan* as it was the only syllable that he could communicate in. Tan was 21 when he was admitted to the Bicêtre hospital (https://psych-classics.yorku.ca/index.htm); Broca found as a result of a lesion, the patient had lost the ability to produce speech. This was only possible to be discovered post-mortem and led to great support for the localization of functions to certain areas of the brain. Later, it would be Carl Wernicke who would find similar results from a lesion that would go on to affect the language production area of the brain. Lateralization would also invite several rebukes to it. This was especially affected by the advent of neuroplasticity. A landmark in the understanding of lateralization and how important it was for both sides of the brain to communicate from Robert Sperry and his split-brain experiments. The brain, as we are aware, is split into two even halves. These two hemispheres are said to be contralateral, i.e., corpus callosotomy was viewed as the solution to reduce epileptic activity and to restrict them to a single hemisphere. These were limiting procedures.

Descartes: The Mind–Body Problem

René Descartes is most commonly associated with his famous words "Cogito Ergo Sum," which translates to "I think, therefore I am." He was a famous French philosopher who was intrigued by the way the body functioned as well. Descartes was among the first to believe that there was a distinction between what was the "mind" and the "body." For Descartes, what was said to be the essence of a person tended to reside within the mind, while the body was merely a vehicle for the mind. These were two entities that were distinct from each other and would go on to influence much of his later theories as well.

Along with the dualism theory, he was the first to propose that the human body contained animal spirits. The animal spirits were a great source of energy for the body and were, as such, the reason that the body was able to function the way it does. The ancient Greeks and other philosophers before Descartes argued that the heart was the center of all activity. Descartes believed that the animal spirits that traveled within pipes in the body originated from the pineal gland. The pineal gland is the connection between the soul and the body and once again reflects the influence of the mind–body dualism theory. He got the idea for this from watching animatronics which were on display in Paris at the time. He associated the wires to the tubes within the body and the animatronic fuel to the animal spirits. In order to avoid the problem altogether, there was also a group of people called the monists. They considered that there was no distinction between the mind and the body and considered them as one and the same.

Luria: Functional Model of the Brain

Luria proposed the three-functional systems model, which stated that the brain areas were functionally wired together (Téllez & Sánchez, 2016). These are specialized areas that carry out particular roles in the brain. However, these units alone are not likely to carry out the roles themselves and that may require them to integrate themselves with other similar units. The brain as a whole functions as a result of this coordination and cooperation between the various areas. These often result in the various behaviors that the brain is capable of as well. Many of the cognitive functions like attention, perception, language, and memory also require several brain areas to cooperate and also make contributions in their own right (Andrewes, 2015).

His model also goes on to describe the various units of the brain that interact to provide a basis for these functional systems. These are the units for regulating tone and waking and mental states, the unit for receiving, analyzing, and storing information, and finally, the unit for programming, regulation, and the verification of activity. The first unit is associated with the functional system of the reticular activating system that is responsible for the states of consciousness along with other functions such as sleep and arousal. The second unit is associated with the areas that deal with the incoming information from the various sensory organs. For example, the occipital lobe (vision), temporal lobe (auditory), and the parietal lobe (somatosensory). The last unit is mostly associated with the frontal lobe. This area is responsible for executive functioning and also controls the functions of the rest of the brain as well.

Neuropsychology in India

The history of psychology in India can be traced to the establishment of the Indian Psychoanalytical Institute in Calcutta at the beginning of the 20th century. The Britishers were influenced by the works of Sigmund Freud and other psychoanalysts of the time, and this was reflected in the psychology that was practiced as well. The subject continued to be the sole domain of the Westerners, and it did not have any basis in the country it was being practiced. However, the subject spread from Calcutta, and other prominent institutes of the time were offering courses in clinical psychology at the bachelor and post-graduate levels. The All India Institute of Mental Health (AIIMH), currently known as the National Institute of Mental Health and Neurosciences (NIMHANS), Bengaluru, saw its inception in the year 1954.

However, the focus of these programs was on the diagnosis and management of mental illnesses. The study of the brain and the assessment of various cognitive domains were restricted to clinical psychologists. The tools used continued to be a product of Western countries. The problem with these was that they were not reliable or even valid measures of the patients in the subcontinent. It was under the initiative of Professor C. R. Mukundan in 1975 that led to the establishment of neuropsychology in India. Professor Mukundan drew inspiration from Alexander Luria, a prominent Russian neuropsychologist, who has also been deemed the "Father of Neuropsychological Assessment." All of Luria's methods and assessments were then adopted to suit the Indian population. The NIMHANS Neuropsychological Battery was also a product of this adaptation. The use of electroencephalography and event-related potentials ensured that clinical and research work could be done in the department (Kumar & Sadasivan, 2016). Researchers continued to work in the area of rehabilitation and intervention methods in neuropsychology for treating impairments and for healthy functioning. With improved access to resources and collaboration with international experts in the field, neuropsychology gained momentum in India. The field saw the development of new tools for assessing cognitive functioning in cases of clinical and neurological impairments. The tools such as "NIMHANS Neuropsychological Battery for Elderly" and "NIMHANS Neuropsychological Battery for Adults" were designed to cater to the unique needs of the Indian population. In recent years there has been a massive increase in the use of advanced research methods, such as neuroimaging and virtual reality, for clinical and applied behavioral research.

Neuropsychological Assessment

Neuropsychology uses a variety of assessment methods to understand behavior and study the brain parts and functions governing those. Methods give us insight into the dynamics, functionality, structure, connectivity within brain parts, normal or aberrant growth within the brain, and the connection of all these with human behavior. Assessment methods support diagnosis, intervention, and rehabilitation as necessary. Neuropsychological assessment includes neuropsychological batteries for a range of cognitive functions, individual tests for a particular sphere of cognition, and tests to ascertain the behavioral and psychological effects of aberrant brain deficits such as dysfunctions in body movement, language processing, understanding others' perspective (Theory of Mind), coping, memory, and emotions.

In this chapter, the available methods are briefly reviewed and will be then expanded with their application in Chapter 2. Standardized neuropsychological tests are the most traditional methods of assessment where the tasks designed are such that they capture specific neurocognitive processes during the performance of the tasks – for example, the Wechsler Memory Scale, Boston Naming Test. Brain scans were used to understand the structure and function of the brain. It was used to assess brain injury by providing images with higher resolution and examining different brain areas. Further, with the advancement in technology, global brain projects are used where brain models based on mouse or monkey and theoretical neuroscience concepts like working memory and attention are developed, which enables mapping brain activities. Electrophysiology is the other method where using electrophysiological measures the brain activity is measured by measuring the electrical and magnetic field. Certain experimental tasks are designed to measure the reaction time of specific neurocognitive processes, for example, CANTAB (Cambridge et al.).

The new methods of investigation involve functional magnetic resonance imaging (fMRI), positron emission tomography (PET), magnetoencephalography

(MEG), optical imaging, and transcranial magnetic stimulation. These techniques overcome the challenges faced using traditional methods and provide better resolution and clarity to the brain structure and functions. As they are non-invasive and require minimal cooperation from patients, their efficiency of usage is comparatively more and better (Lassonde et al., 2006).

Researchers have also mapped various neural activities in the brain through brain scans that involve the usage of software products. They are effectively informative to understand the application of neuropsychology in behaviors. For example, Fooya is a mobile app that studies the influence of dietary preferences among children.

Assessment is considered an essential aspect of any surgeries preceding brain surgeries. Initially, the Wada test was used for these purposes, where X-ray machines were used to take pictures of the flow of the dye through the arteries. The medicine is injected to anesthetize the brain in specific locations while testing functions on the other side of the brain. Memory functions are also assessed similarly. However, this has several disadvantages, as they are uncomfortable and highly invasive. But with the introduction of new methods of investigation, such issues have been controlled.

Research Perspective

This section focuses on the highlights of growth and research in various subfields of neuropsychology in the past decade, presents the key challenges and limitations from multiple perspectives in neuropsychological research, and informs about the future direction of neuropsychological research with advancement in technology, methods of assessment, and better understanding of brain functions.

Every sub-field in neuropsychology has been progressing in its own way, but when we take a comprehensive look, it is observed that the contemporary approach to practice is multifaceted. There has been a paradigm shift in studying cognitive functions and mental disorders (psychiatric and neurodevelopmental). A holistic approach has emerged in recent decades in studying the intricate association between cognitive functions and mental instead of exploring them in isolation. This research direction has seen an upsurge in studies acknowledging the interdependent and reciprocal influence of neurocognitive functions and neuropsychiatric disorders, for example studies on visuospatial working memory in dementia and Alzheimer's disorder, executive functions in major depressive disorder and anxiety, and psychiatric disorders post-traumatic brain injury. Such studies have presented substantial arguments supporting the notion that a holistic approach captures the nuanced dynamics of their interaction.

The field of neuropsychology in recent decades has also seen growth in different countries in terms of research, development, availability of intervention, and a surge in rehabilitation for neuropsychological dysfunction. Studies have identified several challenges when studying cognition and other neuropsychological concepts in varying setups. Though the field has seen progress, it has been variable in different parts of the world. There is a large gap between the requirement and practicing clinical psychologists and clinical neuropsychologists in different countries. Some of the other significant challenges involve the lack of large-scale studies in neuropsychological domains, lack of culturally and linguistically appropriate assessment tools, availability of rehabilitation and intervention techniques to all sections of society, inappropriately translated assessment tools, and lack of well-validated assessment methods (Ponsford, 2017). Packer and Cole (2022) identified seven challenges while conducting a cross-cultural study. The challenges identified are "defining culture, finding representative samples, defining cognition, task variation, ecological validity, interpretation of results, and conducting ethical research." Another avenue of research in this direction is cross-cultural neuropsychological studies, which focus on the effect of cultural differences on neuropsychological practices, how cultural differences shape our brain and neuropsychological performance, and why it is crucial to develop culturally specific tools for assessment, intervention, and rehabilitation.

Rapid progress in clinical neuropsychology practice also comes with its own set of challenges. Some of the challenges include a lack of standard definitions of the domains (constructs) being assessed, a clearly defined set of ethical considerations during clinical practice and research, the impact technical advances are posing on neuropsychological practice, and not enough focus on exploring biosocial causes driving behavior dysfunction (Kent, 2020).

Digital neuropsychology is another domain of research that has seen unprecedented growth in the last couple of decades. It is a developing field that comes with its own unique set of advances and challenges (Germine et al., 2019). Benefits of using digital methods in neuropsychological practice include more straightforward regular assessment of patients in their natural environment, improved ecological validity of data collected, easier development and implementation of personalized intervention plans for patients, and accessibility of services to each section of society. The challenges include the development of norms to validate the application of digital tools for assessment, the cost, and applicability of digital tools are still in the critical stage, and the lack of large-scale studies to assess the implication of such tools for different neuropsychological conditions and how will that impact the neuropsychological practice in future.

Neuropsychological research and clinical practice also include ethical constraints. Some of the prominent challenges include a lack of information on neuropsychological disorders, which poses a stigma in most of the population, a lack of facilities and remuneration for practitioners, a lack of accredited practitioners in the field, and ethical issues to be considered while implementing new technologies for assessment and rehabilitation.

The future scope of neuropsychological research encompasses a wide range of topics in the development of neuropsychological assessment, implementation of new technology at different steps of management of neuropsychological dysfunction, development of efficient intervention techniques and rehabilitation plans, curating the plans for every patient's need, biomarkers for early diagnosis of disorders such Alzheimer's disorder, neurodevelopmental disorders in children, and neural network underlying various cognitive dysfunction such as emotion regulation, working memory, and episodic memory.

In subsequent chapters as we delve into different domains of neuropsychology, we will further explore the current research direction, challenges, and future scope in each domain while understanding the intricacies of neural network underlying behavior and cognitive processes.

Book Plan

This book aspires to be readily accessible for students while not compromising on the academic rigor that would help a professor or a researcher use this as a reference text. The book will be divided into three units, each dealing with a particular aspect of the subject. Part 1 is an introduction to the subject and a brief refresher for those familiar with it. It shall follow a similar theme to the first chapter, where a historical and chronological approach will be taken. The following chapters will deal with the anatomy of the brain and the methods used to study them. It is necessary that the user of this book is well-versed in the structure and functions of the brain as they will come into play in the book frequently.

Part 2 deals with various domains that are also the functions of the brain. These cognitive domains are more likely to be utilized. These will be arranged in order of complexity. This represents the processes as they occur within the brain and are also arranged in their logical sequence as and when the brain encounters any stimulus. Some areas that will be covered are attention, memory, and language.

Part 3 deals with disorders that have a biological basis. These are more likely to affect the brain and directly impact the behavioral aspects. The structural, as well as the corresponding functional deficits, along with current research, shall also be discussed. Some of the disorders that shall be covered are traumatic brain injury, memory disorders such as amnesia, and neurodegenerative disorders like Alzheimer's dementia. Along with these disorders of language and communication, attention and executive dysfunction shall also be reviewed.

The main focus of the book in every chapter that follows shall be to incorporate new research while also recapitulating what has already been uncovered so far. This would open up new areas to be explored along with any future directions that research can be taken in. Lastly, this is among the first books on this subject from the subcontinent. Therefore, special emphasis shall be given to the cultural aspects that are unique to this geographical area.

References

Andrewes, D. (2015). *Neuropsychology: From theory to practice*. Psychology Press.

Benton, A. (1988). Neuropsychology: Past, present and future. In F. Boller & J. Grafman (Eds.), *Handbook of neuropsychology* (Vol. 1, pp. 3–27). Elsevier Science.

Benton, A. (2000). *Exploring the history of neuropsychology: Selected papers*. Oxford University Press.

Benton, A. L., & Sivan, A. B. (2007). Clinical neuropsychology: A brief history. *Disease-a-Month, 53*(3), 142–147. https://doi.org/10.1016/j.disamonth.2007.04.003

Berlucchi, G. (2009). Neuropsychology: Theoretical basis. In L. R. Squire (Ed.), *Encyclopedia of neuroscience* (pp. 1001–1006). Academic Press. https://doi.org/10.1016/B978-008045046-9.00996-7

Carus, P. (1905). Conception of the soul and the belief in resurrection among the Egyptians (illustrated). *The Monist, 15*(3), 409–428.

Cubelli, R., & De Bastiani, P. (2011). 150 years after Leborgne: Why is Paul Broca so important in the history of neuropsychology? *Cortex; A Journal Devoted to the Study of the Nervous System and Behavior, 47*(2), 146–147. https://doi.org/10.1016/j.cortex.2010.11.004

Finger, S. (1994). History of neuropsychology. In E. C. Carterette & M. P. Friedman (Eds.), *Handbook of perception and cognition* (2nd Edition, pp. 1–28). Elsevier.

Finger, S. (2000). *Minds behind the brain: A history of the pioneers and their discoveries*. Oxford University Press.

Germine, L., Reinecke, K., & Chaytor, N. S. (2019). Digital neuropsychology: Challenges and opportunities at the intersection of science and software. *The Clinical Neuropsychologist, 33*(2), 271–286.

Goldstein, G. (1990). Behavioral neuropsychology. In A. S. Bellack, M. Hersen, & A. E. Kazdin (Eds.), *International handbook of behavior modification and therapy* (pp. 139–149). Springer US. https://doi.org/10.1007/978-1-4613-0523-1_7

Humaida, I. A. I. (2019). The psychology of the brain and behavior: Their mutual impact. *American Journal of Psychology and Behavioral Sciences, 6*(3), Article 3.

Kent, P. L. (2020). Evolution of clinical neuropsychology: Four challenges. *Applied Neuropsychology: Adult, 27*(2), 121–133.

Kumar, J. K., & Sadasivan, A. (2016). Neuropsychology in India. *The Clinical Neuropsychologist, 30*(8), 1252–1266. https://doi.org/10.1080/13854046.2016.1197314

Lassonde, M., Sauerwein, H. C., Gallagher, A., Thériault, M., & Lepore, F. (2006). Neuropsychology: Traditional and new methods of investigation. *Epilepsia, 47*(s2), 9–13. https://doi.org/10.1111/j.1528-1167.2006.00680.x

Meier, M. (1974). Some challenges for clinical neuropsychology. In R. Reitan & L. Davison (Eds.), *Clinical neuropsychology: Current status and applications* (pp. 289–323). Wiley.

Mora-Bermúdez, F., Badsha, F., Kanton, S., Camp, J. G., Vernot, B., Köhler, K., Voigt, B., Okita, K., Maricic, T., He, Z., Lachmann, R., Pääbo, S., Treutlein, B., & Huttner, W. B. (2016). Differences and similarities between human and chimpanzee neural progenitors during cerebral cortex development. *ELife, 5*, e18683. https://doi.org/10.7554/eLife.18683

Packer, M. J., & Cole, M. (2022). The challenges to the study of cultural variation in cognition. *Review of Philosophy and Psychology, 14*, 515–537. https://doi.org/10.1007/s13164-022-00637-x

Ponsford, J. (2017). International growth of neuropsychology. *Neuropsychology*, *31*(8), 921–933. https://doi.org/10.1037/neu0000415

Stebbins, G. T. (2007). Chapter 27—Neuropsychological testing. In C. G. Goetz (Ed.), *Textbook of clinical neurology* (3rd ed., pp. 539–557). W.B. Saunders. https://doi.org/10.1016/B978-141603618-0.10027-X

Téllez, A., & Sánchez, T. J. (2016). Luria's model of the functional units of the brain and the neuropsychology of dreaming. *Psychology in Russia: State of the Art*, *9*(4), 80–93. https://doi.org/10.11621/pir.2016.0407

2 Methods in Neuropsychology

Introduction

Neuropsychology is about exploring the intricacies of the brain and unraveling the dynamics between the brain and behavior. Exploration of the realm of neuropsychology requires versatile, robust, and comprehensive methodology through which every nuance of brain organization and its relationship with behavior and vice versa can be studied. The brain has been the subject of much scrutiny for several centuries at this point. It is an organ that is quite essential for the survival of humans. The only method to observe an organ in such a case would only be possible after the human being has died. While this has also yielded a lot of information and knowledge that has expanded the horizons of what the brain is capable of, the real challenge was to observe the brain as it functions in real time. An organ that is responsible for keeping one of the most complex creatures on the planet is a fascinating subject that has kept neuroscientists busy. The methods to study the brain have also followed the progression that was outlined here. The brain is a mass of tissues and cells, and there are few discernible markings on it. The anatomical methods were applied to cadaver brains to highlight and study individual areas and understand their functions. As technology kept advancing, there was more. Prior to phrenologists pointing to the cortex as the seat of mental activities, it was unclear which parts of the brain, when damaged, caused the formation of the deficit.

Neurological Assessments and Methods

Neuro-Histological Procedure

The neuro-histological procedure enables the study of tissues, their structures, their genetic makeup, etc. This is the study to identify, characterize, and describe the cell's functions and dysfunctions. The procedure involves several steps. Initially, the tissue is taken and preserved in 10% formalin or 4% paraformaldehyde and then is kept in 30% sucrose solution with phosphate buffer saline to maintain isotonicity and cell structure and prevent the formation of ice crystals. This enables fixation by preventing bacterial degradation and helps

DOI: 10.4324/9781032640839-3

keep the cell in its lively condition. In the next step, the tissue is hydrated and then dehydrated in different concentrations of alcohol. Then, the clarification procedure is done by keeping it in chloroform for 24 hours to remove the alcohol. Then, it is immersed in paraffin wax, which is melted at 58°C so that it hardens the tissues. This process is called wax impregnation. After impregnation, the hardened block is sliced using a microtome, where wrinkled sections are obtained. To enable clear visibility, the wrinkled wax section is placed in a water bath at 50°C, and then a slide is submerged, and the section is picked up by draining out excess water. Then it is placed in an oven at 37°C for better adherence. Here, either albumin or gelatin that has chromium is added to the slides to ensure better adherence to the slide. Then the slide is placed in Xylene for 15 minutes. This process is called dewaxing. Here, it becomes hydrated. Further, it is dipped in different concentrations of alcohol to dehydrate. Then, it is placed in distilled water and is taken for staining. After staining, it is dipped in different concentrations of alcohol, followed by immersing it in xylene, where a drop of mountant like Distyrene P-Phthalate Xylene is added and finally covered using a coverslip (Slaoui & Fiette, 2011).

Anatomical Method

There are several histological procedures when it comes to studying the brain's anatomical structure. It isn't possible to understand the brain by just looking at it. It is, however, possible to take as thin slices as possible to study it under microscopes. The most important step is to halt the degeneration of the tissues. There are autolytic enzymes in the tissue that contribute to its decomposition at the time of death. The use of fixatives such as formalin is the method to stop this process. The next step in the process is to use staining. By using certain stains, it becomes possible to highlight certain parts of the brain. The basic types of stains used are for the cell body, myelin stains, and degenerating axon stains (Alturkistani et al., 2016). The next step in the process is called dehydration, where the excess moisture is removed from the tissues. This helps solidify the tissues and helps it become more solid to be cut. Then the tissue is embedded into paraffin wax to help get easier extraction of the cellular structures. This is then "sectioned" into ribbons that are then mounted on microscope slides for examination. Modern histological techniques make use of stains that are suited to different areas of the neurons. Masson's stain is used for connective tissues, the Golgi stain is used for neuronal fibers, and the periodic acid-Schiff stain is used for carbohydrates.

Degeneration Method

As the name suggests, these methods study the different parts of the brain and certain neurons by destroying them. The lack of a certain structure is more likely to illustrate the functionality that is lost. This draws a direct connection between the structure and function of that particular area. This is similar to ablation (Latin "to take away"), where that particular part of the brain is removed in order to study the outcome in functionality.

Lesion Method

Lesions are another method that is similar to degeneration. But it differs in that the damage is more precise and can be localized to specific brain regions. Treating the functional damage in that area would indicate brain functionality as well. By understanding what is missing, it would become easier to study the normal function of that area as well. Lesion techniques enable studying the relationships between the brain and behavior (Lavond & Steinmetz, 2003). It involves seven different ways of causing lesions: aspiration lesion, radio-frequency lesion, knife cuts, cryogenic lockage, excitotoxic lesion, sham lesion, and bilateral or unilateral lesion. In aspiration lesions, some parts of the brain are removed, especially the parts behind the eyes, and their effect on behavior is studied. In radio-frequency lesions, the lesion is usually done in the subcortical regions using high-frequency radio waves. The intensity and duration of the radio waves determine the effects of the behavioral changes. In knife cuts, a small knife-like structure is inserted through the skull. The knife is usually closed; once it reaches the targeted location, it opens up to cause the lesion usually used for cortex regions. In cryogenic lockage, a coolant is injected into the brain using microinjections until the targeted neuron or part of the brain is implicated and behavioral changes can be noticed. This is a reversible lesion technique as the neural pathway is blocked for some duration of time, and then, the coolant is removed. In excitotoxic lesions, radio waves are used that implicate the other regions along with the targeted region in the brain. In a sham lesion, a small knife's structure is passed through, which damages the other regions along with the targeted region. In bilateral and unilateral lesion techniques, any of the methods can be used to cause lesions either in one or both hemispheres.

Stereotaxic Method

Stereotactic surgery began as a method in 1908. This was also around the time when the *zeitgeist* of specific areas of the brain being connected with specific functions began to emerge. There was a need to develop a surgical technique that would have the capability of targeting these areas with a level of precision. Victor Horsley, a neurophysiologist and neurosurgeon, along with Robert Clarke, a mathematician, were the first to conceptualize this method based on the 3D Cartesian coordinate system. This was the stereotaxis where stereo means "3D" and taxis means "orderly arrangement." The apparatus was developed through collaboration with Canadian neuroanatomist Aubrey Mussen before it was deemed to be fit for human testing and experimentation. However, this device was never really used for the same and was found discarded at Mussen's apartment in London. While Horsley and Clarke are widely hailed as the inventors of stereotactic surgery, it was actually a combined effort of researchers from across the globe.

The basic idea behind the stereotaxy was based on the idea of localized structures for the different structures of the brain. However, most of these devices, as they existed like Kroenlein and Kohler's cephalometry, Broca's radiography, and Kocher's craniometer, were inaccurate. This was largely due to the fact that these depended on the structure of the skull and, as such, did not take into account the variations of these structures in comparison to the external landmarks on the skull. This problem was solved by Spiegel and Wycis, who jointly developed the first stereotactic equipment that was used for human surgery. This led to the rise of radiosurgery, which uses radiation instead of blades to excise unwanted tissues. This was due in part to the accuracy that was afforded by the stereotactic equipment. Leksell (1983) described that radiosurgery was only held back because there was no method to localize the different brain structures. This was also developed through the advent of computed tomography (CT). The working of the CT scan along with other brain imaging techniques shall be discussed in detail later in the chapter.

Wolf et al. (2019) carried out a detailed review of the current status of stereotactic radiosurgery in this decade. When Leksell introduced the concept in 1951, it was hailed as a revolutionary method as a non-invasive surgery method to remove tumors in the brain. The Leksell Gamma Knife has become one of the commonly used equipment in stereotactic radiosurgery. There are also mask fixation techniques that are reportedly less cumbersome than the frames being used.

Stereotaxic surgery is the most minimally invasive technique used to perform actions like lesion, ablation, injection, stimulation, implantation, and radiosurgery. It works on the basis of three major components where the apparatus has three coordinates for the frame of reference for the brain. Here, bregma is taken as the point of reference from the brain atlas, and the route map is fed into the instrument. The aspiration needle follows the direction as that of the atlas and further causes the lesion or damage to the targeted region.

Neuroimaging Methods

In order to understand the workings of the brain, it is necessary to be able to look at it and the various components that it has. This is a non-invasive method of studying the brain that does not do any harm or damage to it. It was also now possible to carry out experiments with it as well. It opened up new areas for research. Most of the techniques under it broadly fall under two categories: structural and functional. The structural imaging techniques involve looking at the structures of the brain and the neurons themselves, while the functional aspects look at the activity within the neurons as well. It enables the study of cognitive functions and helps in looking at any abnormalities like lesions or tumors at a more intricate level.

Contrast Skull X-Rays

Contrast X-rays are different from the X-rays. While X-rays involve passing the X-rays through the body in order to form an image on a sheet, contrast X-rays use a dye to achieve the same. The dye is called a contrast medium that helps organs and tissues absorb the rays to form the image. Bones usually have a higher density and as such they absorb the radiation easily, however, the organs can only show up with the help of the dye. While this will not give a very detailed picture, it will help identify any tumors in the brain as it will help understand its size and location. However, this has some risks as there is exposure to radiation, and it is not advisable for repeated use.

Computed Tomography Scan Method

Computed Tomography scan was introduced in the early 1970s as an improvement on contrast X-rays. This method involves taking multiple X-rays from various directions, all of which are taken and compiled into cross-sections of different brain parts. While in the chamber, the patient is surrounded by the rotating X-ray tube. CT scans are still relevant due to the fact that they are quick, accurate, and affordable, however, there have been concerns raised with regard to the exposure to radiation. CT images are used for diagnosis of brain injuries and any other conditions related to the brain. CT scans have also been used in surgeries as part of stereotactic radiosurgery for tumors or any other malformations in the brain. It is limited as it presents a static image of the brain and does not give any indication of the changes that may happen within the brain at the moment of the scans itself.

Electroencephalography Method

Electroencephalography (EEG) is a method to record electrical activity inside the brain. It is a non-invasive procedure as the electrodes are placed on the scalp itself. The electrode records the current fluctuations within the brain with regard to a single reference point. The EEG can be used for clinical as well as for research. The spontaneous electrical activity gets recorded over time, and this is where EEG really shines. It has greater accuracy with respect to time; however, the brain signals give limited information about functioning. This is most commonly used to diagnose cases of epilepsy. Epilepsy is sudden, abnormal electrical activity within the neurons, which often disrupts EEG readings, leading to easy identification and diagnosis. Hans Berger has been credited with recording the first human EEG in the year 1924 and, as such, has been widely acknowledged as the inventor of the electroencephalogram as well. This was hailed as a landmark in the history of clinical neurology.

In medical usage, it remains the main diagnostic test for epilepsy. A typical recording takes about 20–30 minutes which also includes recording time. The scalp is cleaned and prepared, after which electrodes are placed with the help of a paste or a gel. These are to enhance the conductivity of the electrodes.

Apart from epilepsy, EEG is also used in diagnosing various sleep disorders, stroke, and encephalitis (inflammation). For research, it is primarily used in conjunction with event-related potentials (ERPs) in the field of cognitive neuroscience, neurolinguistics, and psychophysiology.

Magnetoencephalography

Magnetoencephalography is similar to the EEG. However, instead of the electrical activity, it measures the magnetic activity. Any sort of electrical activity is bound to produce magnetic fields as well as stated by the law of electromagnetic induction. These are so faint that they can only be picked up by superconducting quantum interference devices (SQUIDs). It has excellent temporal resolution and can isolate areas up to a centimeter. An advantage over the EEG is that it is less likely to face interference from the surrounding tissues.

Magnetic Resonance Imaging and Functional Magnetic Resonance Imaging

Magnetic resonance imaging (MRI) uses a chamber that has a highly charged magnetic field. The MRI images are constructed by the interaction of the waves with the waves emitted by hydrogen atoms. Unlike the EEG, the MRI has a great spatial resolution which enables clear differentiation of tissues and other brain parts. It is also relatively safer as it does not use X-rays or any kind of radioactive substances. An fMRI is done to ascertain the functioning of the brain and its various parts. This level of activity is measured using blood oxygen levels in various areas. The more oxygen is used in an area, the more that area is active. While the patient is in the MRI, there are given tasks that will activate the brain areas so that it shows up on the screen as well.

Positron Emission Tomography

Positron emission tomography is a method that uses emissions from radioactively labeled substances. These radiotracers are used to produce positron-emitting radioisotopes when they are injected into the bloodstream. These then produce two- or three-dimensional images that provide a functional picture of the brain and particularly the neurotransmitters. These also allow images of spontaneous activity, however, this might not be as quick as EEG. It primarily works on the basis of glucose metabolism and blood flow in the tissues themselves. This still continues to be used for diagnosis as well, however as the radiation tends to decay quickly, it cannot be used for a long time.

Virtual Reality Systems

Virtual reality (VR) was first introduced as an entertainment system that was nothing more than an expensive toy. Since then, there has been a lot of development of the same, and it eventually became cheaper and more easily accessible. At its essential, VR is an interface between humans and machines that go beyond the monitors. They develop a 3D environment that can be used in a variety of scenarios. Unlike traditional 3D simulations, what sets VR aside is the fact that the digitally rendered environment is fully interactive as well. This is done with the help of handheld controllers and a head-mounted display (HMD) that resembles a mask. The tracking system also simultaneously simulates the motion of the body within the rendered environment.

It has the potential to be a sound system and an easy source of research and assessment in various fields, including neuropsychology. The main issue with most interventions in neuropsychology is that it lacks ecological validity. Most of the interventions are administered in clinics or labs, where the conditions are entirely under the control of the clinician and are typically a situation that is not likely to exist outside of it. In a VR setting, the conditions are still very much under the control of the experimenter or the clinician but with the added advantage of making the situation as realistic as possible. It has also been proven that these are quite similar and accurate simulations of the real world (Schultheis et al., 2002). This means that each of the experiments in the VR setting can be tailored to the individual while still maintaining consistency between the participants. It might be too early to use just VR for assessments, and hence, this could be used as a supplement to the traditional paper-pencil tests and behavioral observations. This could improve the reliability and validity of the measures and would also lead to easier replication of the results.

There are specially designed HMDs and VR equipment that tailors to the deficits that may arise in patients with traumatic brain injury (TBI), multiple sclerosis (MS), and stroke. There are also VR scenarios that incorporate the Wisconsin Card Sorting Test (WCST) in a format where instead of selecting a card, the participants had to walk through a door. These scenarios are called virtual environments (VE), and these can be customized accordingly for most scenarios.

While these are considerably better options, it still needs to be determined whether it is cost-effective. There are many different options, but the utility of a VR system comes with its own disadvantages as well. There is something known as simulator sickness that tends to happen. It is quite similar to motion sickness, and there has been evidence that it affects participants with MS and TBI. The availability of VR also makes it a technology that is not really relevant to use just yet.

Brain Electrical Oscillations Signature from India

When it comes to solving crimes and aiding investigations, neuroscience and neuropsychology have made great strides. This can be seen in the importance given to forensic evidence in a case. These are the proofs that are often accepted in a court of law and allow the lawyers to prosecute the guilty efficiently as well. These have been popularized in many detective television series and movies as well. Most cases usually use a method that has already been discussed in this chapter, the polygraph. In a similar strain, a contribution from India to the fledgling field of forensic sciences is the Brain Electrical Oscillations Signature (BEOS).

BEOS Profiling is the latest in advanced methods of criminal investigation. It uses a multi-channel EEG that is attached to the suspect, and they are given no other stimulus or expected to respond in any way. The suspect is interviewed without the BEOS and then later made to sit in a room that is empty except for a screen. After the EEG is attached to the suspect, the screen shows them a list of probable events or scenarios that are likely to have happened in the crime. These are also read out as audio probes or even include visual probes. These are shown sequentially in the form of a list. The suspect merely looks at the screen. At the end of this procedure, it would be easy to know if the suspect really did commit the crime.

In order to better understand the basis behind BEOS, we need to adequately distinguish between two cognitive processes. These are knowing and remembering. Both of these processes are connected to long-term memory but work in slightly different ways. Knowing about something merely indicates that you might have at some point in your life read or memorized or even encountered the fact. Suppose we view the same in the crime scene. It could be argued that a person or even the suspect might know about the details of the case based on newspaper reports. However, this is something that is usually picked up by the polygraph. BEOS looks for instances of remembering the facts and scenarios as they might have happened. In our daily lives, we are constantly retrieving stored information from our brains. This means that when we are remembering something, we are more likely to retrieve that information in the long term. A person can only remember something if they have personally experienced it. Remembrance is related to autobiographical memory or personal experiences only. When it comes to the personal experiences that an individual has, these consist of both sensory-motor and emotional participation in the event. If you experience sunrise at a hill station, your senses and your motor pathways are active and the event also evokes a certain emotion. Both the sensory information and the emotion are now bound to that memory of the sunset. When you recollect or remember that same memory, it leads to the reactivation of the same brain areas that were active during the actual experience. Although there have already been sentencing made on the basis of information from BEOS, its credibility in the research community is still shaky at best, and its use has been cautioned in further cases without the validation of its methods.

References

Alturkistani, H. A., Tashkandi, F. M., & Mohammedsaleh, Z. M. (2016). Histological stains: A literature review and case study. *Global Journal of Health Science, 8*(3), 72.

Lavond, D. G., & Steinmetz, J. E. Eds. (2003). Lesion techniques for behavioral experiments. In *Handbook of classical conditioning* (pp. 249–276). Springer US. https://doi.org/10.1007/978-1-4615-0263-0_8

Leksell, L. (1983). Stereotactic radiosurgery. *Journal of Neurology, Neurosurgery & Psychiatry, 46*(9), 797–803.

Schultheis, M. T., Himelstein, J., & Rizzo, A. A. (2002). Virtual reality and neuropsychology: Upgrading the current tools. *The Journal of Head Trauma Rehabilitation, 17*(5), 378–394.

Slaoui, M., & Fiette, L. (2011). Histopathology procedures: From tissue sampling to histopathological evaluation. In J.-C. Gautier (Ed.), *Drug safety evaluation: Methods and protocols* (pp. 69–82). Humana Press. https://doi.org/10.1007/978-1-60761-849-2_4

Wolf, A., Naylor, K., Tam, M., Habibi, A., Novotny, J., Liščák, R., Martinez-Moreno, N., Martinez-Alvarez, R., Sisterson, N., & Golfinos, J. G. (2019). Risk of radiation-associated intracranial malignancy after stereotactic radiosurgery: A retrospective, multicentre, cohort study. *The Lancet Oncology, 20*(1), 159–164.

3 The Philosophy of Mind, Consciousness, and Cognition

Introduction

Philosophy is a mixed bag of emotions when it comes to it. At its best, it becomes a question of asking too many questions. At its worst, there will always be questions that have no ready solutions at all. However, it is the questions that usually shed light on finding more answers. They tend to open up a whole new perspective that hadn't been explored up until that point.

In order to understand the true nature of the mind, it is essential to question it at every step of the way. Consciousness is a state of being aware of one's surroundings and being able to respond to them in kind. We are only able to be aware of our surroundings because we can *sense* them. Therefore, a certain level of evolved neurobiology is necessary before one can be conscious. But then, does it mean that only animals who can *sense* or *feel* as we define it are conscious? Does an insect then not possess this quality? Would defining consciousness on the basis of how we experience it not lend a touch of subjectivity to an objective definition? These are questions that have plagued philosophers, to neurobiologists and might also haunt AI designers in the future.

There are three significant domains when it comes to understanding consciousness: the philosophical, the neurobiological, and the neuroevolutionary of the subject. The philosophers were the first to arrive on the scene and ponder what it means to exist, sense, or feel. Only through questioning everything can one come in proximity to the truth. René Descartes, the French philosopher, was the first to deliberate about the nature of the brain and the existence of a mind that was independent of the brain. However, the biggest problem posited by philosophers was that if consciousness was a *subjective* experience that was unique to the individual, would there really be an *objective* scientific theory that would explain it? Here is where the neurobiologists took over and tried to simplify the domain of consciousness through their time-honored technique of *reductionism*. Reducing any phenomenon to the electrochemical activity of the neurons has proven beneficial to the advancement of neurobiology in scientific circles. We know that epilepsy is a result of the abnormal electrical activity in certain neurons that we sleep because the brain releases melatonin. However, there is still no ascertained neurobiological mechanism

DOI: 10.4324/9781032640839-4

for consciousness in the brain. The neuroevolutionary perspective has given a criterion for what could be defined as consciousness. While the long-standing view was that brain development and intellectual development were necessary, this would mean that only humans are truly conscious. By that logic, if consciousness were to be determined in that sense, a future where machines also gain consciousness would not be too far off. We are surrounded by them, and much of our behavior lies on the continuum between whether something is an object or a being that is capable of *feeling*.

No one knows where or when consciousness arises. It raises more questions than it answers, and there has been a lot of speculation with no end in sight. Humans are blessed, or some would argue cursed by it, which has ultimately driven us as a race to understand ourselves better. There are many concerns when it comes to consciousness in other species. However, it is important to understand where this truly comes from. Epithalamus is the seat of consciousness in the brain.

If we take a look at what the Indians thought of consciousness, we will find that there was indeed no area of philosophy that was left untouched, and in his article titled "Sri Aurobindo's essentially Vedic concept of Consciousness," Cornelissen (2008) argued that his concept did not differ much from the already existing Vedic ontology on the same. Consciousness, as outlined in the Vedas, has been outlined to include the entire universe as a whole as well.

Indian psychology has a perspective that is unmatched by its contemporaries. It looks at the transcendental experiences that humans have. These are then used as a medium to explain everything. Indian psychology also does not shy away from the concept of past life regression either, acknowledging the often overlooked aspect in other regions. In Hindu philosophy, the concept of *Advaita* and *Advaita* is often used to differentiate between reality and illusions (*Maya*). The Advaita philosophy believes that the world is accurate and that everything around us is also real. It is also a dualist perspective where they believe that the soul and the God, or the divine creator, are different. There is more of a sense of devotion and an expectation of protection. The Advaita philosophy doesn't believe in these differences and instead believes that the soul is merely an extension of God.

The entire composition of the universe is made of three qualities, the *trigunas*. These are Sattva (balance), Rajas (activity), and Tamas (inertia). Sattva is the state of balance between the other two and is essential for happiness, wisdom, and self-control. Rajas are usually symbolic of activity without the sattva aspects and overactivity. In a similar sense, Tamas looks at how these would stop and halt progress in terms of inactivity.

There are typically five approaches when it comes to Eastern philosophies: Vedas, Upanishads, schools like the Samakhya, Buddhism, and Jainism. These explain consciousness as having four levels, according to the Upanishads. These are Jagruta, which is the awareness level; Swapna, which is the dreams; Sushupti, which is a trance-like state that is similar to daydreaming. A fourth factor that is completely different from the rest is Turiya that relates to the universe itself. The Jains believe that there is only one level of consciousness and that this is only possible with the body and does not exist without it.

When Asian philosophy looks at either of these three concepts, there can be no doubt that there will be a melting pot of all the major religions as well as regions involved in it. In order to understand the overall perspective of Asian or Eastern philosophy, it becomes necessary to evaluate each of the dominant religions and regions that make up the Eastern side of the world. When it comes to understanding the entire Asian or Eastern philosophy, evaluating it one at a time is usually more straightforward. In this case, looking at the Indian perspective would be a good start. In particular, Buddhist philosophy played a significant role in the development of the movement in India. Siddhartha Gautama, an emperor who sacrificed his powerful position to live as a monk, was a pivotal figure in the expansion of Buddhism as a philosophy in the region. The general agreement that persists in Indian philosophy is that the mind is not the brain, and neither is it just the medium of processing information. It has been conceptualized as a cognitive process that comprises successive mental states. Each of these can be broadly categorized as "cognizing" or learning and "being aware" (Buddh). There is a general consensus that exists with regard to the extra-physical aspect of the mind. It is something that cannot be felt or touched. However, there are certain exceptions to this where the ability of an entity does not necessarily exist as independent of the entity itself. To simplify this with an example, the very fact that water quenches thirst is a property of water and not something that becomes the quenching property of it. Buddhism is unique in that it disregards the self and tries to put forth the idea of a collective or the universal rather than the individual. Like the Vedas of the Hindus, the Buddhists followed the teachings of the Abhidharma. These are ancient texts which are revisions of the words that appeared in the original sutras. The self and the mind exist independently of each other. The mental activity requires the cooperation of the two fundamentals that would make up the universe (Dreyfus & Thompson, 2007).

Western Perspective/Philosophy of Mind

The Western perspective seeks to elaborate on whether the two entities, namely, the mind and the brain exist independently of each other. The mentality or the mental states that exist may either be based on the brain itself, i.e., has a physical basis to itself, which is a tradition that is known as Naturalism. The other explanation is that mental states are a product of themselves and do not rely on anything to justify their existence.

Naturalists believe that the mind is a part of nature. This is, by definition, Monism which is held in sharp contrast to French philosopher René Descartes' Dualism. The Naturalists believe that everything in existence has a natural, physical property to it, whereas Descartes believed that most things, including humans, were compositions of the physical and the non-physical entities as well. There also exist Cartesian souls that will exist irrespective of the existence of the physical such as a brain or the body. Naturalism holds the belief that most things that exist are physical, and these all come from nature. This is known as Emergent Naturalism. It regards the properties that might exist and that these properties are a part of the entity and would be meaningless without the existence of the entity itself. It is important, at this point, to differentiate between properties and attributes. While a property might be something that might not function without a basis, it would be different from an attribute. A block of ice is cold, and this is a property that has been assigned to the union of ice. However, when we speak of attributes, they are bound to be non-physical such as thinking or feeling.

There are two varieties of Emergent Naturalism: Nomic and Moorean Emergentism. Nomic Emergentism posits that there are certain fundamental laws that link the physical properties of an entity to its mental aspects. These fundamental laws are as immutable as they are in natural sciences. There are connections between the physical and the mental because the laws say that there are these connections. David Chalmers is a proponent of this type of philosophy. On the other hand, Moorean Emergentism, as given by G. E. Moore, is the assertion that there is a fundamental dependency on natural sources and moral goodness. The phenomenon and the physical are fundamentally the same. This distinction and dependency are metaphysically necessary and continue to exist despite the absence of a law.

There has also been a clamoring for change from this view, which seeks to explain the philosophy outside of these connections that have been assumed so far. In accordance with this *zeitgeist*, there has been a shift away from Emergentism. The mental states do not emerge, but in keeping with the Naturalistic tradition is material. In the case of such a philosophy, the mental and physical states tend to be one and the same. Therefore there is no real difference between the two. This is usually countered by the fact that there may also be beings who do not share this construct, and the physical may not necessarily hold the same way it does for human beings. This brings us back to Supervenience Materialism which asserts that physical and mental states tend to be dependent on each other and that such a relationship of dependence can also be explained rather than just being stated as laws.

The real problem in the philosophy of mind can usually be traced to the difficulty of differentiating between the physical and the mental states or materials. Most physical states or materials are reducible to their basic forms. The mental states or forms are irreducible and, therefore, would not be able to explain using the same principles as the ones used for physical states. This is where functionalism stepped in. It sought to identify these entities using the functions of the properties. These functions would have to be realized and ascertained as a whole and, therefore, required to be experienced. Would having the mere knowledge of experience be the same as having experienced the same? This is a question that seems to be straightforward at first, but there is no simple answer to it. Without experiencing something firsthand, can you truly claim to understand it wholeheartedly?

References

Cornelissen, M. (2008). Sri Aurobindo's evolutionary ontology of consciousness. In H. Wautischer (Ed.), *Ontology of consciousness: Percipient action* (pp. 11–52). MIT.
Dreyfus, G., & Thompson, E. (2007). *Philosophical theories of consciousness: Asian perspectives.* Cambridge University Press.

4   Neurons and Structure of the Nervous System

Introduction

There are many organs within living beings. That being said, there is one that is uniquely linked to the human body that is the brain. Human beings have the most advanced brain on the planet. This is primarily due to the development of the frontal lobe that is responsible for the higher-order thinking skills and the executive functioning. More importantly, it forms the links between all the other parts of the body as well. There is one primary center in the body where all these wires meet and are then sent out to the other. The brain, the spinal cord, and the nerves are the communication system. In our technologically rich society, perhaps it would be easier to illustrate this relation with the Internet. The brain is the main server or service provider that is responsible for the exchange of data. The other parts of the body are the various users who would like to talk or interact with each other. The wires that connect these users (the body still functions without wireless technology, unfortunately) are the nerves. This would give an accurate idea when we go on to talk about the three broad divisions of the nervous system.

The central nervous system (CNS), peripheral nervous system (PNS), and the autonomic nervous system (ANS) are the three broad divisions of the nervous system as a whole. The CNS comprises of the brain and the spinal cord, the PNS is the connection of the body with the CNS and the other parts of the body. All the information about the body like the senses of touch, smell, pressure, sight, and hearing are communicated to the CNS via the PNS. The ANS is a specialized circuit that is a combination of the CNS and the PNS. It primarily overlooks, as the name suggests, the autonomous or automatic functions within the body. These are important for the survival of the human body. Some of these functions include breathing, pumping of the blood, digestion, and sexual arousal. These are all functions of the body that do not require any voluntary action on the part of the individual. The functions and the structure of the ANS would be discussed in better detail later in the book.

When it comes to the brain, it is important to understand that each and every part of it is vital to its understanding. It looks like a huge mass of tissues that has many folds know as sulci and smooth parts known as gyri. Apart from

DOI: 10.4324/9781032640839-5

these, the brain is also divided into many sections and parts so that it would be easier to help identify and refer to its many areas. Any specialized field is filled with jargons, and while these might act as gatekeepers to fully understanding that subject, and may make it difficult, its knowledge is also a means of identification of a person's proficiency in it. Most of these terms are in Latin, yet they are the widely accepted versions for the same.

As shown in Figure 4.1, rostral means toward the head, while caudal means toward the tail. Dorsal means toward the back, while ventral is toward the stomach. The terms posterior and anterior may also be used to indicate the same. Lateral means the sides, while medial indicates the midline. You may notice that there are about two terms that may mean the same. Rostral, anterior, and ventral all seem to be indicating the front direction. There is no clear indication of how these might be used, so be sure to refer to these whenever these come up. The terms all differ with respect to the position of the head and the eyes, where they might be looking directly up, or straight in front. Another set of terminology that is usually used for the brain is the planes that the brain occupies. Students familiar with the Cartesian coordinates will be quick to confirm that the position of any object that occupies space can be determined using the three planes, x, y, and z. Similarly, for the brain, there is the coronal plane that appears as a vertical cut at the crown. The horizontal plane which cuts the head in its usual posture and the sagittal plane that divides the brain into two halves along the interhemispheric fissure or the medial longitudinal fissure, the deep groove at the midline. The planes are particularly helpful in identifying the regions of the brain during neuroimaging techniques as discussed in the previous chapters.

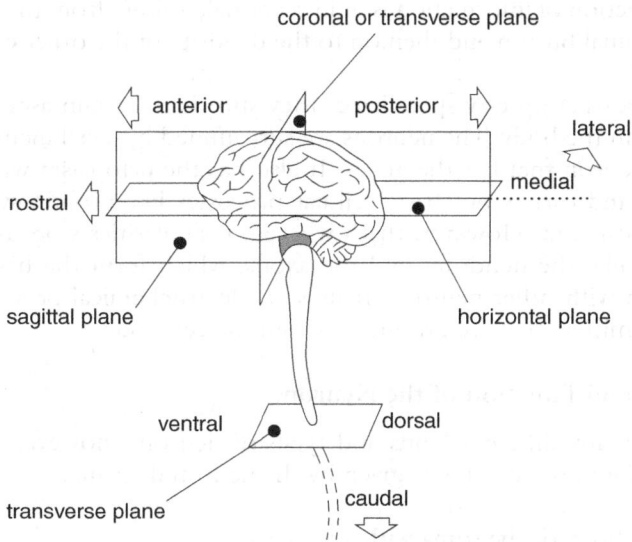

Figure 4.1 Terms used to describe various sections and directions.

Neurons and Glial Cells

When it comes to cell bodies, each of them is anatomically and functionally different in order to fulfill the purpose of helping the overall human body survive and thrive. The neuron is different in that it is solely purposed for transfer of information and communication. Also, while the communication between these cells is important for the overall functioning the body, these do not touch each other and as such is not one continuous system. Before we discuss the structure of the neuron, the basis for the entire neuron system, we must discuss why they become such an important aspect of the history of neuropsychology. Ramon y Cajal has become one of the most widely cited figures in the field of neuroscience for his assertion of the neuron theory or the neuron doctrine which also saw him win the Nobel Prize along with Camillo Golgi. The staining techniques were the key to identifying the whole structure of the neuron. It was believed that the neural network was one continuous structure that did not have any free endings or discontinuous junctions either within itself or whenever it connected to another system like the skeletal or the muscular system. It was Cajal and his series of neuro-histological studies that gave rise to the understanding and conceptualization of the neurons as we know even to this day. The neuron doctrine can be summarized into four major points:

* The neuron is the fundamental structural and functional unit of the nervous system.
* Neurons are discrete cells and do not have any connections with other cells.
* The neuron as a cell is composed of three parts – the dendrites, the axon, and the terminal button.
* The direction of information is unidirectional, it flows from the cell body to the terminal button and then on to the dendrite of the other cell.

While neurons are pretty specialized, they still share certain aspects with the other cells in the body. The neurons are surrounded by a cell membrane, containing a neuron that has the genes. It also has the cytoplasm with the usual organelles and carries out basic cellular processes like protein synthesis and energy production. However, that's where the similarities stop. It has special extensions like the dendrites and the axons, which form the basis for communication with other neurons. It uses an electrochemical process that uses neurotransmitters that are produced within the cell body.

Structure and Function of the Neuron

There are many different kinds and types of neurons, however, all of them have four common features as given by the neuron doctrine.

1 The cell body or the soma with a nucleus
2 Dendrites
3 An axon
4 Terminal synaptic buttons

Cell Body

The neuron is survived by the cell body as it forms a major part of its structural integrity. As these are grey in color, the term grey matter is used to describe the regions of the brain that have a lot of them. The cell body contains the same structure and function as the other cells in the body. This also contributes to the proteins that are present in the axon as well.

Dendrite

The neurons are specialized cells for information processing, and this information only travels in one direction. The dendrites, which are directly connected to the cell body, are where the information sent by other neurons is received. There can be thousands upon thousands of these dendrites connected to a single neuron and tend to differ depending on the function of the neuron. These have been compared to tree-like structures with many branches being attached to the surrounding neurons. Dendrites are covered in spines and these are also the biggest parts of the receptive surface of the neurons

Axon

Axons are the part of the cell that extends away from the cell body and transmits the information to the other neurons. These are variable in length and can be anywhere from about less than 1 mm to about a 1 m. Depending on the type of the neuron, they may also vary in terms of the number of axons that are connected to the neuron as well.

Many of these axons are surrounded by a myelin sheath that is supposed to increase the speed of the electrochemical transmissions. The myelin is a protein that surrounds the axon in layers and gives it the appearance of white matter. The sheath forms an insulation that increases the conduction velocity. This sheath is formed by the glial cells, which are the support cells for the neurons. In the CNS, the oligodendrocytes form the myelin, and the Schwann cells do the same in the PNS. As can be observed in Figure 4.1, there are gaps in the sheath which are known as the nodes of Ranvier. These are a mechanism which serves to further increase the rate of transmission as it jumps between the nodes.

Terminal Buttons

The ends of the axons are slightly enlarged and form ends known as terminal buttons. It forms one end of the synapse which is a gap between two neurons and the site of interneuronal communication. The terminal button releases the neurotransmitter and depending on the type of neurotransmitter released, it might either affect the other neuron in an inhibitory manner or an excitatory manner. This is where the electrochemical aspect of neuron comes into the picture. The message that is sent from the cell body is an electrical impulse, however, the message is then converted into a chemical message in the form of the neurotransmitter and then passed onto the next neuron. This is also the reason why we are able to study neural activity using either electricity-based methods like the electroencephalography (EEG) and evoked potential and metabolic based techniques like the positron emission tomography.

Classification of the Neurons

Neurons are usually divided into various groups on the basis of their structure (morphology) and their functions. Most neurons are defined on the basis of the number of extensions from the cell body. Consequently, a majority of them are classified as multipolar neurons. These would include more than one axon or dendrite. Bipolar neurons are those that have just two of these extensions while unipolar neurons are those that do not have any dendritic extensions. Neurons with short axons are called interneurons and these help in better integration of the neural activity within specific areas of the brain.

Classifying neurons on the basis of their function would give us the two dichotomies of motor and sensory neurons. The motor neurons, also known as efferent neurons, are from the brain to the body and make the muscles move and control functions for other organs as well. The sensory neurons otherwise known as afferent neurons are those where the information (which is unidirectional) goes from the body to the brain, or more accurately the CNS. These carry information about the various senses. A majority of the neurons tend to be interneurons so that these opposing tracts of neurons are connected to each other as well.

In order to adequately influence behavior, there are usually many neurons that are bundled together. Behavioral changes that arise are a result of all of these firing together. These are usually known as tracts or fibers. There are three types of these fibers that appear as white matter within the CNS.

1 Intracerebral (association) connects regions within the same hemisphere.
2 Intercerebral (commissural) connects structure between the two hemispheres.
3 Projection fibers connect subcortical structures to the cortex and vice-versa.

Glial Cells

These outnumber the neurons, however, due to their really small size, they occupy less than 50% of the volume in the brain. Glia comes from the Greek gliak, which means nerve glue and as such functions as the support system for the nervous system. There are multiple functions of the glial cells and as the neurons are differentiated in terms of their functions, so too are the glial cells. These functions involve providing nutrients to the neurons, taking away the waste products, and also destroying any neurons that have been injured. These glial cells are present wherever there are neurons and as such are present throughout the CNS and the PNS. In the CNS, there are four types of glial cells which are oligodendrocytes, astrocytes, microglia, and ependymal cells. In the PNS, there are the Schwann cells.

Astrocytes

They get their name due to the fact that they look like a star (Astra). It has a central cell body that is surrounded by several cytoplasmic projections. It typically sits between a blood capillary and a neuron. It has many tiny dendrites, and these provide support between the dendrites and neurons as well. The astrocyte also functions as the blood brain barrier support squad. It prevents toxins and poisons from entering the brain. It acts like a force field of sorts. These also contribute to the overall structural integrity of the BBB. This in essence keeps any large molecules away from the brain such as drugs, and only allows water, gases, and small, soluble lipids. It regulates the flow of ions, sugars, oxygen, and carbon dioxide into and out of the neurons. These are also responsible for taking away any excess neurotransmitter left after the neurons have fired and also restore the ion levels around the neurons to facilitate in action potential. Their connection with blood capillaries is used to provide the glucose to the neurons in a form that is easily usable. As a student, you might have come across images and models of what the brain actually looks like. This very book contains these and also describes what it really feels like as well. If you are lucky enough to visit a brain museum (there is one in NIMHANS, Bengaluru), you can touch a brain and its many layers. This organ, even in its formaldehyde avatar, is pretty malleable and soft. Such a thing needs multiple layers of protection. These will be discussed in detail later in the chapter.

Ependymal Cells

These are cube-shaped cells that form the lining of the brain's ventricles and the spinal column's central canal. Their main function is the production of the cerebrospinal fluid (CSF). This is a clear liquid that fills internal cavities within the brain and the spinal cord. This also protects the brain and the spinal cord, mechanically and immunologically.

Microglial Cells

It is the smallest of the neuroglial cells and this helps it move around the CNS more easily and efficiently. It is the janitor of the neuroglial cells, and it supports the neurons by phagocytizing (eating up) bacterial cells and dead cell debris.

Oligodendrocytes and the Schwann Cells

These are found in the CNS and the PNS, respectively. Their main function is to provide the myelin sheath, which you may recall provided the insulation around the axons and helped speed up the transmission of the electrical signals from the cell body to the axons. The Schwann cell also provides the same to the axons present in the PNS.

Neurodevelopment – Anatomical and Functional Development of Neurons

The central nervous system forms the major and the most complex part of the nervous system. It tends to develop in segments and as each portion comes into existence, it becomes more and more complicated. The nervous system is the first system to start developing, and it is also the last system to finish developing. This is testament to the complexity and the vitality of the nervous system. Like all other systems within the body, even the nervous system develops according to the two basic principles of development: determination and differentiation. Determination is the mechanism within the cells which ensures that a group of embryonic cells change into the nervous system and no other type of cells. Fetuses are initially just a huge mass of cells, and it is using these initial cells that the rest of the body and the other systems are developed. Once determination has occurred, the corresponding migration, specialization, and connections of these newly formed neurons and glia are called differentiation. The process differentiates these cells as being special and more appropriate to the function as determined. Soon after fertilization, the embryo begins to differentiate and become specialized. As the nervous system is the first to begin developing, the outer layer of the embryo known as the ectoderm is the basis for the formation of the neurons. This breaks away as a plate or sheet of cells that form a long tube-like structure. It is hollow and resembles a pipe. One end of it forms the spinal cord and the other end develops three swellings that form the three major distinct areas of the brain: forebrain, midbrain, and the hindbrain. Each of these areas further divide into their functional and structural areas. As this continues to happen, the brain wall becomes thicker and accumulates more mass. At this stage, the determination has already happened. All the cells have been determined to form parts of the nervous system. Now, the differentiation process begins. As the cells keep maturing, they migrate to the areas they are supposed to as part of their genetic makeup and accordingly form their specialized connections. Cells from the inner part of the tube migrate to the outer parts and form the six layers of the meninges. It develops inside out and this forces the newer cells to push past the older ones to form the newer layers. At this stage, the primitive neural cell is called a neuroblast and it seeks to establish connections around itself. However, this process does not happen at random and is dictated by chemical gradients. These may either repel neurons that are inconsistent with the area or attract the ones so that the appropriate pre- and post-synapses are connected. This also leads to the formation of glioblasts that go on to form oligodendrocytes and astrocytes. These are capable of increasing their numbers all by themselves and in this manner proliferate all over the nervous system. During embryogenesis, the glial cells direct the migration of neurons (Goldstein & McNeil, 2013).

Meninges and the Cerebrospinal Fluid

As discussed earlier, the brain itself is a pretty fragile organ and an equally vital one. Hence, there is no other organ that is as protected in the body. Going for the outermost layers of protection, we come across the skull that encloses the whole of the brain as part of the skeletal system. Next come the meninges, which are layers of covering between the tough skull and the brain. There is also the CSF that is the liquid in which the brain is suspended so that it doesn't suffer any shock damage from being moved around or by impact with the layers designed to protect it. Lastly comes the BBB that is designed to take care of the environment in which the brain functions.

The meninges are the three layers of protective membranes that surround the brain. The outermost layer is called the dura mater. This is made up of two layers: periosteal, which attaches to the skull and the meningeal, which attaches to the next layer. The second layer is the arachnoid layer, which is more of a spider web-like structure. Finally, the pia mater is a soft, delicate membrane that closely adorns the many convolutions of the brain. The spaces in between these meninges are called cranial meningeal spaces. Depending on which space is being referred to with respect to the meninges, they are called "epi" (above) or "sub" (below). Of all these spaces, the subarachnoid space is the one that holds functional utility as it is filled with CSF and also the blood vessels that supply blood to the brain.

The CSF is a clear, colorless fluid that acts as a medium for transportation within the brain. It brings nutrients to the many, many neurons in the brain and also takes away all the waste that they generate. The CSF is produced by the choroid plexus within the lateral, third and fourth ventricles of the brain. As discussed earlier, it functions as a shock absorber for the brain and is a mechanical buffer. The average adult has 140 ml of CSF, and nearly 400 ml is produced every day.

Neural Impulse, Communication among the Neurons

The communication between neurons takes place in the form of electrochemical reactions between the two neurons. While the connections between the neurons work on electricity, it doesn't transmit the same way as you might be familiar within the electrical wires that connect your homes to the station. The communications and its high rate of transmission mean that there has to be the passage of ions through tiny channels in the axon walls. Some of these are sodium ($Na+$), potassium ($K+$), calcium ($Ca2+$), and chloride ($Cl-$). Ions develop an electrical charge whenever they lose or gain electrons. The passage of these ions is the cause of electrical buildup and transmission in the axon. The axon holds a negative charge with respect to its outside. This is a high difference of nearly -70 mV (millivolts). This constant high tension is necessary to require rapid action from the neurons. Most of the sodium ions are usually kept outside despite being attracted to the negative charge within the axon.

Action Potential

The action potential is basically the charge that is required for the neurons to fire. This works on the all-or-none principle. This means that the neuron will either fire or it will fail to fire. There is no in-between when it comes to that. For that to happen, there has to be a depolarization, which in simple terms means that the negative polarity inside the axons has to change to positive. This is facilitated through the semipermeable membrane of the neurons. They allow the rapid passage of the positive sodium ions outside the neurons to enter the axon hillock. This changes the polarity and raises it to 35 mV, which is the amount necessary to start the action potential. Anything below this potential will not cause the neurons to fire as dictated by the all-or-none principle. This nerve impulse spreads along the axon as the voltage-controlled sodium channels open sequentially down the axon. Neurons are continuously firing at all times, however, even they have a small-time interval known as the refractory period during which they will not be able to fire. It is like loading the neuron up and restoring the electrical charge so that it might be able to fire once again. The impulse is transmitted in this manner along the axon until it reaches the end at the terminal buttons. The communication between neurons is in the form of synapses. Depending on the neurons, it might carry two types of neurotransmitters and they either have an excitatory or an inhibitory effect. The cumulative effect of these neurotransmitters is not on the basis of the number of neurotransmitters but the firing rate of these neurons that carry the corresponding neurotransmitter. The location where this communication takes place is known as the synapse. The combination of these synapses sets up the network which are more or less interconnected to each other. The neurons that fire together, wire together, and the more a particular network is used, the more its effects tend to get stronger and faster.

Synapses

Before we get into the technical details of the synapse and the exact process by which the communication takes place, let's digress into a more historical phenomena: smoke signals. Imagine the following scenario. You are stranded on the side of a cliff. Between you and the next cliff is a huge chasm nearly a kilometer wide. On the other side of these chasm however is your friend who has unfortunately been separated from you in this freak accident of nature. You can try shouting or try calling, but none of these seem to work. The only effective way of communication that you both were able to come up with is the smoke signals. Keep in mind that this is very much a hypothetical situation and that all other variables that your overactive minds can think up are safely negated. The point being, in order to make sure that your smoke signal is effectively visible on the other side, you would have to combine multiple fires so that there is enough smoke produced. This is the only way that the urgency or the clarity of the message can be sent to the other side. You and your friend are the neurons and the chasm are the synapse, the smoke that you used are the neurotransmitters. This is a gross oversimplification of one of the most complex procedures that takes place in the brain. But hopefully, it has given you a clarity of what takes place in the brain when we talk of the synapse. That's all the point of the story was.

When it comes to neuronal connections, its medium changes from the electrical to the chemical. The action potential travels from the cell to the buttons. These are the presynaptic neurons which contain buttons that are huge and bulbous so as to increase the area of contact with the postsynaptic neuron; however, they are not in direct contact with the postsynaptic neuron, there is a very tiny gap known as the synaptic cleft that can only be seen by an electron microscope. The buttons have oval structures called vesicles that contain the neurotransmitters that have been synthesized by the cell body. These tend to stay close to the membrane of the presynaptic neurons. The receptor sites for these are situated in the postsynaptic neurons. Each neurotransmitter is a molecule, of a certain shape. This means that each of them also has a receptor site that is uniquely suited to them. This is similar to a lock and key mechanism, so that the expected situation occurs for the expected match between a neurotransmitter and a receptor site. There are two types of receptor sites, one which is symmetrical leads to an inhibitory response and the symmetrical one usually leads to an excitatory response.

When the neural impulse reaches the buttons, each of the vesicles release a certain number of neurotransmitters into the synaptic cleft. In most cases, the neurotransmitters tend to bind with their corresponding receptor sites on the postsynaptic neurons. They blend into the cell membrane and either cause the neuron to fire by changing the ionic permeability of the membrane so that there are increased levels of the sodium and chloride ions in the cell, with the potassium moving out. This essentially depolarizes the cell and creates an excitatory postsynaptic potential (EPSP). This ensures that the postsynaptic cell will fire. An inhibitory exchange would usually result in the membrane being more permeable to the potassium (K+) ions. This causes a hyperpolarization and creates an inhibitory postsynaptic potential (IPSP) that makes the likelihood of the postsynaptic cell firing pretty low. These are the equivalent of yes/no responses that the neuron can give so that the next neuron either fires or it does not fire. This is the main principle of neural communication. However, the point to be remembered is that a single neuron has many synaptic terminals so it could influence many others.

Organization of the Nervous System

The nervous system is usually divided into two: the central nervous system and the peripheral nervous system. These are broad divisions on the basis of the functions that each of these neurons play. There are also anatomical differences in the neurons that exist within these two systems, but these are relative to the functions of the same. The CNS contains the brain and the spinal cord while the PNS contains anything that is not contained by the CNS. The CNS consists of the somatic nervous system (SNS) that interacts with the environs outside while the ANS interacts and regulates the internal systems of the body. The ANS is further divided into the sympathetic and the parasympathetic nervous system. Both these systems are in constant communication with each other and are necessary to remain in perfect harmony.

The PNS and all its components are like the messengers of the CNS. They carry information about the body and its environment to the CNS and then deliver any messages that are required to the body by the CNS. The somatic system has afferent nerves that carry information from the sense organs to the CNS and the efferent nerves that carry motor signals from the brain to the muscles. Most of these nerves are interfaced through the spinal cord, however, there are 12 pairs of nerves called the cranial nerves that directly connect to the brain. The ANS is necessary to control internal organs like the heart and the intestines. It also has the vital fight or flight response in the event of any major trauma or distress. The sympathetic nervous system activates the body to expend energy and increase blood flow and pressure along with the heart rate. The contrary system in this case is the parasympathetic system that reduces the activity caused by the sympathetic nervous system. These two branches of the ANS are required to balance each other out, but they are also largely outside of voluntary control. While this division would be required to help us understand the nervous system, there are many overlaps between these two and these are not always easy to differentiate as we have done here.

The Central Nervous System

As discussed in the chapter, there are three parts of the brain that are developed from the neural tube. There are many major parts of the brain but the primary ones are the division of the cerebrum into the four lobes namely the frontal, temporal, parietal, and the occipital lobes. Apart from these divisions, the brain is also divided into the two hemispheres. Each of these hemispheres act contralaterally, which is to say that the left hemisphere controls the right side of the body and the right hemisphere controls the right side of the brain. These hemispheres are not equal in the functionality afforded to them. This is called the lateralization of functions to the different hemispheres. Broadly speaking, the left hemisphere takes control of language, logic, and positive emotions. It deals with verbal and sequential events. The right hemisphere deals with the spatial and emotional aspects along with the big picture area. This also means that there is cerebral dominance but ensures that these are communicative with each other. That being said, these all act together. In order to better understand these divisions, we will also integrate the functions along with them so as to integrate the two together. The functional areas of the cerebral cortex are divided into the motor areas, the sensory areas, and the association areas that integrate the two of these.

The majority of the frontal lobe has the areas for the motor areas and the higher-order functioning like emotion regulation and planning which are the executive functioning. It has the primary motor cortex, premotor cortex, the frontal eye field, Broca's area, and the prefrontal cortex.

The primary motor cortex is located in the precentral gyrus and is composed of pyramidal cells. These are large neurons whose axons make up the corticospinal tracts. It allows the conscious control of voluntary movements along the skeletal muscles. This is demonstrated in the motor homunculus that is a caricature of relative amounts of cortical tissue devoted to each motor function. There are large areas of grey matter that are relegated to the control of hands, face, and mouth.

The premotor cortex is located anterior to the precentral gyrus and it controls more complex movements like the fingers. These are the learned controlled movements that are repetitive like typing and playing a musical instrument. It coordinates simultaneous actions or any movements that are carried out in sequences. It is also involved in the planning of movements.

Broca's area is located in the left cerebral hemisphere and it primarily deals with speech production. It has a motor speech area that directs the muscles of the tongue. This area is also active whenever someone is preparing to speak. There is a corresponding region in the right cerebral hemisphere that attaches emotional overtones to the words spoken as produced in the region. The majority of the language areas surround the lateral sulcus in the left hemisphere. Along with the aforementioned Broca's area, there is also the Wernicke's area that is involved in the comprehension of speech.

The prefrontal cortex is located in the anterior portion of the frontal lobe. It performs several cognitive functions that involve intellect, cognition, and facets of the personality. It is necessary for judgment, reasoning, and a conscience. This area is also closely linked with the limbic system that is the emotional part of the brain.

Sensory Areas and the Remaining Lobes of the Brain

There are several cortical areas involved in the conscious awareness of sensation. These are located in the parietal, temporal, and occipital lobes. There are distinct areas for each of the major senses in the body. These are the primary somatosensory cortex, somatosensory association cortex, visual and auditory areas, and finally, the olfactory, gustatory, and the vestibular cortices.

The primary somatosensory cortex is located in the post-central gyrus. This is involved with the conscious awareness of general somatic senses that receives information from the skin and skeletal muscles. This projection is contralateral which means that it receives sensory input from the opposite sides of the body. It also has a somatosensory homunculus that is similar to the motor homunculus.

The somatosensory association cortex is located posterior to the primary somatosensory cortex. It integrates the sensory information received from the various senses in the body. The association area makes a comprehensive understanding of the stimulus as reported by the sensory organs. In a similar sense, the visual areas comprise of the primary visual cortex. This is located on the posterior part of the occipital lobe and receives visual information from the retinas. This works in tandem with the visual association area. This surrounds the primary visual cortex and it interprets the visual stimuli like color, form, and movement.

The auditory areas are similarly designed like the visual areas. They have the primary auditory cortex that is located at the superior temporal lobe. This receives information related to pitch, rhythm, and loudness. The auditory association area is located posterior to the primary auditory cortex and it stores memories of the sounds and permits perception of these sounds as they are stored in the brain. It is also involved in recognizing and understanding speech as it lies in the center of the Wernicke's area.

As a support for all these sensory areas, there is the association areas that essentially integrate all these different sensory areas. They associate any new sensory input with experiences. This has also led these to be termed as higher-order processing areas, which includes the prefrontal cortex and the language areas.

Brainstem and the Cerebellum

The brainstem and the cerebellum are the older parts of the brain. These are the areas that are evolutionarily old and the first parts of the brain that were already present with respect to the new brain or the cerebrum. These two areas are responsible for the functions that are essential for our survival and are purely autonomous. The brainstem includes the medulla oblongata, the pons, and the structures of the midbrain. These are present in almost all organisms with the same structure. The medulla is immediately superior to the spinal cord and forms a zone between the spinal cord and the complex system of the brain. It is also the point where the contralaterality of the brain kicks in. The pons are basically two bulb-like structures inferior to the midbrain. The midbrain is between the cerebrum and the pons. It is the smallest part of the brainstem. The brainstem is the connection between all the major sensory and motor tracts to and from the body.

In terms of the function that this part plays in the body, there are many. As discussed earlier, it plays reflexive functions that are necessary for life along with auditory processing, visually guided movements, and the control of movements. The medulla takes care of respiration, blood pressure, and heart rate. These are all reflexive and do not need continuous voluntary inputs. The pons also helps in balance of the body and is in charge of our sense of directions. Similarly, the cerebellum controls the coordinated movements and is also involved in thinking and language.

Regeneration of the Neurons

Most parts of the body when under duress or injury are likely to repair themselves so that there is close to no loss of functionality barring a few minutes or hours depending on the severity of the injury. However, the same can't be said for the nervous system. Whenever there is any trauma to the central nervous system, it is expected that three types of repairs must occur. The first is the axonal regeneration or regrowth, the second is the restoration to the damaged neurons itself, and finally, the regeneration of neurons to replace the damaged ones.

The main complication arises in the fact that the very complexity of the neurons makes it so that they are harder to repair. This inevitably leads to loss of functionality that is pretty permanent. This means that lower organisms that boast of simpler neural networks are more likely to recover from any damage than the higher-level organisms such as humans. This does not mean that it is impossible but that it just tends to be more complicated. Even in humans, the repair to the nervous system is more likely to happen in the PNS rather than the CNS. This does not mean that there is any fundamental difference in the structures of the neurons itself but rather a nod toward the often-ignored part of neuropsychology: the glial cells. The support cells as discussed in this chapter are different for the CNS and the PNS. While there are many of them in the CNS, there is only one, namely the Schwann cell in the PNS. The Schwann cell not only provides the myelination for the axons in the PNS but it also provides an environment that is conducive to the repair and regeneration of damaged axons. These cells are not limited to just producing myelin but also produce extracellular matrix proteins like laminin and fibronectin that are essential for axonal growth. They also provide neural cell adhesion molecules that act as glue for restoring parts of the axon. Finally, they contribute to the neurotrophic factors that restore neurons. These are the three mechanisms that the Schwann cell deploys in the event of any injury in the PNS. Also, it must be noted that most of these repair mechanisms are limited to the axons and the repair is easier since the Schwann cells remain on the axons in the PNS. Any CNS neurons with their axons in the PNS are also likely to be repaired easily due to the Schwann cells. These differences are due to the differences in the oligodendrocytes and the Schwann cells. The Schwann cells are more likely to proliferate and grow themselves so that they can repair in the event of a trauma but the oligodendrocytes do not react the same way. Most of the CNS is designed to limit the extent of damage caused by the trauma and therefore its priority is to contain the damage first rather than to repair it completely. The glial cells in the CNS like the microglia are prone to go apoptotic than the ones in the PNS (Steward et al., 2013).

The more traditional methods that might come to mind seeing these differences in the CNS and the PNS would be that if the hypothetical environment was the major issue, would the use of the conducive environment in the PNS be better. So and Aguayo (1985) demonstrated this in rats. They grafted PNS cells into the damaged retina of the rats. It showed that these grafts from the PNS were able to help in the axonal growth of the damaged retinal nerve. This was also established as proof–of-principle that it was not that the neurons are incapable of regeneration but that the environment of the CNS itself is not conducive to promote such a regrowth. One of the main reasons for this harsh environment is due to the presence of the astrocytes. The astrocytes are another class of glial cells that help in providing structure and nutrition to the neurons. Their secondary function is that in the event of any trauma to the CNS, they release chondroitin sulphate proteoglycans. These proteins are responsible for the formation of glial cells in the CNS. It was found that these tend to remain in the CNS for a lot longer than the formation of the glial scar and hinder the process of axonal regrowth as well. Apart from these, there are also the NOGO proteins that are also responsible for the inhibition of the axonal regrowth of the damaged axons in the CNS. Circumventing the mechanisms of either of these proteins has shown that it might be possible to encourage the regrowth of axons on the CNS as well.

Alternative Methods for Cell Regeneration

These are all methods that can be used to enhance neuroregeneration. However, these would only work in the event that there is a small-scale trauma. In cases where there is a requirement of large-scale cell replacement, this would not work out. Stem cells are vastly popular in such cases and have become an increasingly important area of research. This is primarily due to the fact that they are self-repairing and proliferative and can also give rise to different types of cells. This is to say that they are not fixed in their differentiative ability and can be used to give rise to many different types of cells as necessary. In this case, adult stem cells are more likely to be the closest solution. These neural stem cells are found in the hippocampus, particularly in the subgranular zone of the dentate gyrus. This region is filled with proliferative stem cells as it plays an important role in learning and forming new memories. The other area that has a similar composition is the subventricular region in rodents. Apart from these, there are also the fetal brain, the spinal cord, and the neural retina. These are often used in vitro and are made to proliferate in the affected region. Once the proliferation inducing proteins are taken away, these stem cells begin to differentiate into the major types of cells like the neurons, the astrocytes, and the oligodendrocytes. A downside to this method is that they reduce the neurogenic property of the stem cells itself and cause it to run out sooner. Hence, most of the focus has been on the neuroregenerative and neuroprotective factors of the stem cells rather than view them as replacement for the neurons. This can then be used for treating various disorders like amyotrophic lateral sclerosis, Parkinson's disorder, and age-related macular degeneration (Steward et al., 2013).

References

Goldstein, L. H., & McNeil, J. E. (2013). *Clinical neuropsychology: A practical guide to assessment and management for clinicians*. Wiley Online Library.

So, K.-F., & Aguayo, A. (1985). Lengthy regrowth of cut axons from ganglion cells after peripheral nerve transplantation into the retina of adult rats. *Brain Research*, *328*(2), 349–354.

Steward, M. M., Sridhar, A., & Meyer, J. S. (2013). Neural regeneration. In E. Heber-Katz & D. L. Stocum (Eds.), *New perspectives in regeneration* (pp. 163–191). Springer. https://doi.org/10.1007/82_2012_302

5 Forensic Psychology – Scope, Challenges, and Applications

Introduction

As a field of research and science, psychology has had to give its pound of flesh as a field that has more to offer than mere commonsensical analysis. The very fact that the preceding sentence is an oxymoron did not dissuade the people from not acknowledging the field and its contributions to improvements in everyday lives. Its contributions highlight the growth and relevance of psychology to the law and crime.

Forensic psychology is the field of psychology that directly deals with the legal and criminal aspects of the justice system. When it comes to forensic psychology, there are two major types that it deals with here. The first aspect is the legal aspect, where they directly apply to the working of the legal system and the courtroom. It deals with the efficacy of evidence, witnesses, and the courtroom. The second aspect is the criminological side of things that is related to the crime and the criminals themselves. Hugo Munsterberg initiated the earliest contribution of psychology to the courtroom in his book *"On the Witness Stand"* (1908). The book dealt with subject matters such as the memory of the witness on the stand, fake confessions, and the susceptibility of the witnesses to suggestions. He posited that it would be unfair to take the word of the witness as being wholly accurate and went on to expound on various reasons as to why such blind faith might result in a miscarriage of justice (The book can be found on archive.org if you are interested in reading it). However, this book did not make the impact he wanted, and much of his contribution to this field disappeared upon his death. It was not until the Devlin Enquiry in 1976 that there was a resurgence in the inaccuracies found in witness testimonies (Davies & Beech, 2017).

Among these budding attempts at establishing forensic psychology as a field of scientific inquiry, it would be incomplete if we did not hint at the earliest ideas. The first name that appears in the discussion of forensic psychology is Cesare Lombroso, an Italian criminologist. He was the first to broach the topic of the heritability of criminal behavior. He was a strong advocate of physiognomy, which was the idea that human traits reflect in the physical, like the body or the face. He meant that a criminal looked a certain way, and

DOI: 10.4324/9781032640839-6

anyone who looked that way was more likely to commit a crime. Some characteristics that he said were criminal were a twisted nose, excessive cheekbones,
long arms, massive jaws and chin, and an almost ape-like appearance. While
this may be considered absurd and borderline racist today, this theory was very
influential at the time. It meant that the criminals did not voluntarily commit
crimes but that they were born that way. This theory, which is eerily similar to
Francis Galton's eugenics, was soon put to rest by Charles Goring. From the
environmental perspective, John Bowlby of Attachment Theory fame believed
that if the child is separated during the second six months from the mother,
it will cause severe damage to its development and well-being. In a similar
strain, many behaviorists like Skinner held the view that such sort of criminal
behavior could be a result of social learning and due to being exposed to it at
an early age. It was also due to improper conditioning, where the bad behavior
was most likely encouraged, and the proper behavior was largely ignored. The
advent of cognitive theories meant that there was also a certain amount of
agency for the criminals themselves to act and that they were not at the mercy
of their circumstances.

In a similar light is the rational choice theory. This theory came out of
the work on situational crime prevention (Wortley & Mazerolle, 2009). The
environmental perspective views the environment as a factor in crime. The
offender and the offense itself are more likely to be influenced by the environment in which it happens. Some examples are using street lights in dark areas
at night, cameras at shops or ATMs, and more patrols. A critique of situational
crime prevention suggested that altering the environment itself was likely to
reduce the occurrence of crime in that area but then move it somewhere else.
The offenders are people who have made a rational choice to commit the crime
and do so to benefit from it as well. There are several stages in the planning
of a crime, and there are many choices that are made in each of these stages.
These are based on the opportunities, costs, and benefits for each that make
the chances of these being displaced to other areas. Hence, any perceived loss
of crime would be rationalized by the would-be offender rather than it spill-
ing over to another area. Committing a crime is not similar to satisfying some
basic need, and it would not harm the offender for not having committed a
crime (Cornish & Clarke, 1987).

The field began to contribute to the law and the courtrooms by bringing
into question one of the most cited pieces of evidence in any legal case: eyewitness testimonies. These are among the most valuable pieces of evidence
that can be presented in a case, actual contact with the crime, and criminal
identification. However, these have come increasingly under question as in
the United States alone, nearly 350 people who were convicted based on eyewitness testimonies were later exonerated by the DNA evidence at the scene
(Albright, 2017). The history of the recognition of the implications of forensic
psychology was a hard-fought one. Gudjonsson (2003) goes on to expand
upon the long, arduous road that forensic psychologists had to face to bring
credibility to a science that was long held illegitimate. One of the significant

challenges that had to be overcome was the blind faith that many prosecutors, judges, and juries would place in eyewitness identification. There was no doubt in their mind that these were wholly accurate. It was considered that only people with mental illnesses would willingly lie in court and under oath. It was also that the credibility of psychologists was under question, and they were not considered expert witnesses. Anytime that they were present in court, they were subjected to ridicule and not taken seriously at all. It was only after a detailed conceptualization of a framework to dispute confessions and testimonies that there was some evidence and credence to expert statements by psychologists (Gudjonsson & MacKeith, 1988).

Crime and Violence

Biological Causes

While the view that psychopathy or the tendency to commit crimes was innate in criminals was disproved since Lombroso's time, there is still an emphasis on finding particular neurobiological markers that could hint at the predisposition to commit crimes. These markers would aid in preventing them as well. A study that was conducted on twins showed that monozygotic twins were more likely to report psychopathic traits, which is an essential link to the genetic contribution to psychopathy (Blonigen et al., 2003). A study with functional magnetic resonance imaging identified that in men who were violent offenders and also been diagnosed with antisocial personality disorder and psychopathy, there was a significant reduction in grey matter volume in the anterior rostral prefrontal cortex and the temporal poles. These areas have been implicated in moral reasoning and prosocial emotions like guilt or embarrassment (Gregory et al., 2012). This means that while no person is born predisposed to commit crimes, there are significant brain differences that could be identified as a biomarker. The offenders do not see anything wrong in the violence, and the traditional controls that society might place on people, such as the connections, do not exist for them.

The frontal lobe is the area that deals with executive functioning and response inhibition. Any damage to that area can also be linked to the crime. Chow (2000) highlighted that any lesions on the frontal lobe could resemble any combination of three frontal-subcortical syndromes. These are the orbitofrontal, anterior cingulate cortex, and dorsolateral prefrontal regions. The most common of these is the orbitofrontal syndrome. Any damage or lesions in this region leads to emotional dysregulation, impaired judgment, indecisiveness, and tactlessness. They are more likely to be outspoken and brash in social situations and not plan things. Patients who manifest anterior cingulate syndrome are prone to be withdrawn and apathetic.

In contrast, the dorsolateral prefrontal syndrome leads to disorganization in their interaction with the environment, and they cannot plan. However, the damage due to trauma or lesions is often diffused and may end up damaging these circuits regardless of any direct injury or lesions in these areas. This difficulty in narrowing down the affected region propounds the difficulty in diagnosing the cause of any personality changes too. These deficits lead to disinhibition in controlling behavior that would otherwise be unacceptable in society. These are usually behaviors that become the norm during adolescence. Many changes happen during this crucial period that is somewhat socially acceptable at that age. An increase in hormones, particularly testosterone, which has been implicated in aggression, means that more teenagers engage in rebellious acts. This period also corresponds with a maturing prefrontal cortex, especially the dorsolateral prefrontal cortex that matures into adulthood. It is necessary to inhibit these aggressive and impulsive urges and mature into adulthood. Damage to these areas means that aggressive and violent behavior tends to carry on into adulthood and even escalate. This significant cognitive load is not quickly processed with the damaged prefrontal cortex. This damage could be both a result of head trauma or late maturity of the cortex. In the former case, this might not be easily rectified, while the latter would be a matter of time (Raine, 2002a).

Other areas that are also implicated in aggression and antisocial behavior are the temporal lobe, the amygdala, and the insular. Dysfunction of this would lead to a reduction in stimulus-response learning. There is no effect of any rewards or, more importantly, the punishments that might have deterred violent behavior (Umbach et al., 2015).

Neuropsychological assessment can also provide a window of predictability when it comes to psychopathy and violence. Many of the batteries include tests for executive functioning and planning. People with antisocial personality are also more likely to have low verbal IQ. There is a reduction in their executive functioning as well, which leads to a reduction in their sensitivity to punishment and increases their tendency to be rewarded.

Hormones

Hormones are large-scale chemical messengers in the body that initiate changes wherever they go. The link between testosterone and aggression has been widely accepted; however, this can only be directly correlated in adults (Raine, 2002b). Pope et al. (2000) demonstrated in a randomized, placebo-controlled trial that testosterone administration to a non-clinical population of males significantly increased ratings on mania and self-reports of uncharacteristic aggression. Nevertheless, this effect was not uniform across all participants. There has not been a lot done on the effect of environmental influences on this relationship. There is indeed a biosocial factor to it. The environment can influence testosterone and other hormones, such as cortisol. It was found that levels of testosterone would increase in men with high socioeconomic status (Mazur & Booth, 1999). The inclusion of honor cultures and strict hierarchies within organizations and societies induce competition among men. Much of what constitutes antisocial behavior often stems from a need or, more aptly, a desire to dominate an authority figure or the rules or laws that have been established in society. It becomes a primer that prepares men for a competition that may escalate to violence. The same was not found in women; however, the positive correlation between status and testosterone levels remained (Mazur & Booth, 1998).

In a study by Dabbs and Morris (1990), it was shown that there was a link between violence and high levels of testosterone. In general, men with high levels of testosterone tend to be delinquent and act in excess in various aspects of social life. Levels of testosterone are linked to dominance and competition, which can lead to both antisocial and prosocial behavior as well. Dominance and competition sometimes get a bad rap, but these are essential and appreciated in many spheres of professional and even personal life as well. It might be a result of the hunter-gatherer society where resources were scarce and only the competent were more likely to survive. It is still pretty prevalent even today in our hypercompetitive society. Instead, rather than the scarcity of resources, it is the access to resources that is scarce. Even in the classroom that you are currently studying in right now, there were perhaps thousands of applications that were rejected to allow the 60 or so of you in. In the study, the participants' socioeconomic status was determined by their education, occupation, and living conditions. While there were participants with high levels of testosterone, it was only the participants in the lower SES group that were likely to be violent or commit crimes. Being able to afford an education means that a person is better equipped to deal with the antisocial tendencies that may arise. Oddly enough, the interaction of biology and antisocial behavior, in this case, was more pronounced in unfavorable environments. This observation contrasts with other markers like low heart rate and low prefrontal activity, which tend to act up in relatively calmer environments. In a similar trend, neurotransmitters like serotonin levels were higher in violent offenders when compared to controls. Those who came from conflicted family backgrounds were three times more likely to commit acts of violence.

Serotonin plays a role in regulating aggressive behavior, and its deficit leads to difficulty controlling violent acts (Moffitt et al., 1997). Research has indicated that the regulation of aggression by neurotransmitters is a complicated process. Out of the many neurotransmitters that are present in the brain, three of them have accumulated the most for their role in aggression. These are serotonin, dopamine, and gamma-aminobutyric acid (GABA). Serotonin is a predominantly inhibitory neurotransmitter. Its effect can be both excitatory or inhibitory depending on the brain region and the receptors that it interacts with. Dopamine also similarly interacts with the mesocorticolimbic system and the reward pathway to enhance aggressive acts. These acts might be a result of a focus on the rewards with little consideration being given to their consequences. Lastly, while GABA is the primary inhibitory neurotransmitter in the brain, it shares a very complicated relationship with aggression.

This relationship is also intermingled with the effect that serotonin might have as well (Narvaes & Martins de Almeida, 2014).

Serotonin

Serotonin levels have been shown to inhibit pathological forms of aggression. When it comes to animal models in studying aggression, a surprising observation is that laboratory rats tend to get domesticated. Using these rats would not yield good results in these studies. Pathological aggression in rats has been categorized as out of control, content, and context. These fall along with the resident-intruder model of aggression, where the behavior of the resident to the intruder is observed. Among the rats, these are indicated by a lack of threatening postures and direct attacks. Another is the continued attack and biting despite the other rat displaying submissive postures – finally, attacks to vulnerable areas or attacking the dominant males, females, or even dead rats. The critical aspect here is that aggression does not always necessarily lead to violence. These acts, as described above, would mean that a certain threshold has been crossed where the intended functionality of aggression has been replaced (De Boer et al., 2009). People are more social than animals, and it is not often that intruder situations occur in everyday life. The more plausible scenario is the economics of living in human society. The ultimatum game is an economic experiment where two participants are involved.

There are a proposer and a responder in this scenario. The proposer is tasked with dividing a certain amount of currency between the two. The responder can only accept or reject this offer. The catch is that if the responder were to reject the offer, neither of the participants would get anything. Healthy participants with altered levels of serotonin would be more likely to reject offers that they perceived as unfair, despite it being a loss scenario (Crockett et al., 2008). When it comes to the sensitivity of neurotransmitters to influence behavior, there are more animal models. It was found that rats that were specifically bred to have short attack latency (these are quick to attack and become aggressive) have more postsynaptic 5-HT1A receptors in the hippocampus. These had an increased binding capacity with the neurotransmitters than long attack latency lines of rats (Veenema & Neumann, 2007). Serotonin is a neurotransmitter that also helps in emotion regulation, and this aspect is essential to inhibiting aggression as well. Serotonin receptor agonists also help regulate levels of aggression, especially with those located in the prefrontal cortex.

Dopamine

Dopamine is a class of neurotransmitters that are a part of the catecholamine family. Its exact role in aggressive behavior is not known yet. Dopamine is associated with aggression and competition. This interpretation and involvement with the reward system makes it self-explanatory and risk-taking as well (Arias-Carrión et al., 2010).

Correctional Method or Interventions

Neurofeedback

Neurofeedback is increasingly being used to tackle many types of disorders in the clinical psychiatric setting, from anxiety disorders to schizophrenia. However, there is no similar trend when it comes to its usage in forensic psychiatry. The presence of a comorbidity is often seen as a contraindication for neurofibrillary tangle (NFT) and is the main reason why it is not frequently used in forensic settings. The forensic psychiatric population includes people who are not wholly responsible for their crimes. Part of the blame lies in the mental illness as well. Fielenbach et al. (2018a) studied the effects of neurofeedback on forensic psychiatric patients, about 19 of them with substance use disorder along with any other comorbidity as well. Primarily, her protocol was to test the ability to reduce impulsivity and restore inhibitory controls within them through increasing sensory-motor rhythm (SMR) and reducing theta. In order to ascertain whether there was any change due to the NFT, the electroencephalography (EEG) magnitudes must have changed from the baseline by about 8%. On the face of it, although there were improvements in their impulsivity and craving due to increasing SMR, there were low respondents. This improvement was attributed to the low number of sessions along with the continued use of their prescriptions as well.

Neurofeedback has opened up the possibility of neurobiological approaches to the field of forensic psychiatry and correctional facilities. At present, it does not enjoy widespread attention from psychologists, primarily due to the overemphasis on psychosocial factors of antisocial behavior and psychopathy. However, traditional forms of psychotherapy are plagued with problems such as low compliance rates and, in multicultural settings, a lack of proper communication. The stage is open for newer approaches, such as neurofeedback. Neurofeedback therapy proved to be instrumental in dealing with cases of aggression and antisocial behavior as a result of substance abuse. It was effective in reducing the craving, as well as abstinence was also prolonged. In cases of domestic violence, which are precipitated by the fear of abandonment and pathological jealousy, neurofeedback was used to enhance flexible thinking and emotional response as well. Similar results were found when the aggression was a result of comorbid disorders such as attention-deficit/hyperactivity disorder (ADHD) and post-traumatic stress disorder (PTSD) (Outsem, 2014).

When it comes to prisoners, there is the aspect of criminal offending, recidivism, reoffending, aggression, violence, and along with disorders associated with the same, like ADHD, schizophrenia, psychosis, cluster B personality disorders, psychopathy, and substance use disorder. An important aspect when it comes to EEG-NFT is that there has to be actual learning for it to work. This criterion takes prime importance when it comes to studies that employ this intervention. A review paper found that out of a total of 224 articles screened, there were only ten that specifically looked at EEG. It included details on the protocol used along with a specific criterion for EEG learning. Impulsivity and difficulties with inhibitory control did seem to improve, along with hostility and drug use. Recidivism could also be reduced if these symptoms are reduced as well. However, it was uncertain whether improvements in brain waves are always correlated with the behavioral measures utilized. The possibility of randomized control trials might help ascertain the effects of neurofeedback (Fielenbach et al., 2018b). The most thought-provoking study of these ten was one which looked to study the aspect of brain self-regulation in criminal psychopaths. It utilized slow cortical potentials demonstrating that the psychopaths were able to control brain excitability. There were reduced aggression, impulsivity, and behavioral tendencies. There are few psychophysiological treatments as intervention, and neurofeedback continues to be the leading candidate for reducing psychopathic characteristics such as disinhibition, aggression, and related behavioral approach tendencies. This observation was also reflected in self-report measures for physical aggression, although reactive aggression and aggression inhibition were not significant (Konicar et al., 2015).

Democratic Therapeutic Communities

People come together for a variety of reasons. There may be a common objective or cause. Sometimes it might be an artificial group that does not last for too long. However, there is strength in numbers that defines existence. Most interventions tend to focus on biological or psychological aspects. However, the social aspects and their interaction with other factors give a complete picture of any disorder.

Similarly, there is a lot to be learned when people interact with others who might be knowledgeable or in a better position to help. Nevertheless, the best way to ensure that information is readily accepted and processed is when a member of the peer group conveys the same. The entire process for which the group is assembled is a lot more interactive and collaborative. These are in contrast with other types of intervention, where there is a one-way street for information that mostly flies clear of its intended recipients. It is the combined effort of the group rather than one individual that pushes for the results. The team becomes an important aspect, and the often-neglected social aspects get remedied in such an environment.

Democratic therapeutic communities (DTC) are community-based interventions that were first introduced for other areas before making their way to the forensic setting as well. The proof of its utility and efficacy lies in the walls of Grendon, a medium-security prison that hosts about 235 men. It is also the first, largest, and a dedicated therapeutic community. These communities are correctly designed, keeping in mind the social environment while offering a program of treatment for its residents. The inmates are called residents in order to lessen the divide between each other and the prison staff. Both of these are equal members of the community, and any decisions that are taken can be made democratically. The traditional prison hierarchy that usually exists is discarded in DTC. These hold psychodynamic theories and a personal development program that aims at reducing the rates of reoffending and recidivism. It is believed that this interaction between the individuals, peer community, and psychodynamic therapy would lead to sufficient insight that would be conducive to long-term changes. These were first introduced during the Second World War for traumatized war veterans who suffered disorders as a result of their involvement in the war. Shell shock or hysteria and other disorders brought about by stress were pretty standard among combat veterans. Many psychodynamically oriented therapists like Adler, Freud, and Mead thought that the general military hospitals were based on authority, hierarchy, and control dynamics. These would only make for a damaging and dependency-prone environment. It was believed that breaking down the traditional roles of the therapist and the client, where the therapist is in a position of power, would be helpful. Instead, emphasizing collaboration and group interaction would be more conducive to healing (Stevens, 2010).

Behavior Modification Techniques

Behavior modification techniques have become the mainstay for many disorders as they are relatively similar to administer and also produce results that can be easily quantifiable. In many forensic settings, aggression toward self and others becomes a significant concern, especially regarding violent offenders or those who have a mental disorder. Previously, interventions such as the token economy and reinforcement schedules have been used with reasonable response rates and success (Comaty et al., 2001; LePage et al., 2003; Oehler et al., 2018). An alternative to this is positive behavioral support (PBS) which is based on the science behind applied behavioral analysis (ABA). Some of the core components of the PBS are (1) ecological strategies, (2) positive programming, (3) focused support strategies, and (4) reactive strategies. These strategies require constant monitoring and readjustment as time and situations keep evolving. Unlike the previously discussed democratic therapeutic communities, there is greater involvement of the staff, and they control the environment without being authoritative. It is an excellent non-pharmacologic intervention to reduce recidivism and reduce rates of reoffending as well. It integrates methods like teaching, modeling, and reinforcing prosocial behaviors based on an assessment of the patients and their environment. It would be particularly helpful for those who remain unresponsive to psychotropic medications and other forms of interventions (Tolisano et al., 2017).

There is collaborative consultation by the treatment team on the behavioral aspects of the patients. The second step is the identification of factors that influence aggressive behavior. Based on the consultation and the analysis, a skills-based program is developed, and this is then handed over to the administration for oversight and monitoring. The central aspect of PBS is the functional behavior assessment (FBA), which tries to understand the meaning or function of the behavior. By understanding the purpose or the motivation behind the behavior, it would be easier to counteract it or modify the environment accordingly. Any PBS plan has the components mentioned above. The ecological strategies remove any mismatches between the individual's needs and their environment. Positive programming or psychosocial rehabilitation involves teaching general social skills and other distress tolerance skills as part of their coping strategy. It also relies on preexisting strategies like the token economy and reinforcement schedules as well. Finally, strategies like active listening and empathy validation would also reduce their aggression as a reaction to perceived slights (Tolisano et al., 2017).

Research Perspective

Forensic neuropsychology has made tremendous progress in the last couple of years with advancements in assessment methods and their applications in the field; for example, the use of eye tracking for eyewitness identification and testimony. Studies have shown that forensic neuropsychology has not only seen progress in research but also in educating clinical neuropsychologists about the specific uses and applications of the field in particular setups by practitioners in legal systems and related settings. In a recent review of the history and current development of forensic neuropsychology, researchers have identified the developmental path of the field, some of the prominent challenges in the field, and its current status (Sweet et al., 2023). Neuropsychological malingering has been one of the critical areas of interest in forensic neuropsychology. It has promoted a wide variety of research in the field, such as the validity of assessment methods. The wide acceptance of symptom validity tests and performance validity tests after extensive research over the years is now proving helpful to clinical neuropsychologists in assessments. Literature indicates that forensic neuropsychology has a wide range of applications, including evaluation of kins in case of property issues, evaluation of mental processing after brain injuries or stroke, in case of an individual being a part of a legal situation, evaluation of elderly people in judicial cases, in the assessment of children and adolescent in unsafe custody, etc.

Therapies that are given to mental health patients in forensic settings are another prominent area of research. The review by (Abbing et al., 2023) has indicated a lack of systematic overview of art therapies such as music therapy, drama therapy, visual therapy, etc. The studies indicate that art therapies have a positive impact on modifying regulatory processes and positive impact on social functioning, psychiatric symptoms, etc. Furthermore, this review presents the limitations and future scope in the application of art therapies. There is a dire need to develop essential assessment tools mainly to be used in forensic setups, a larger sample size for validation of studies, a strong research methodology, and the development and testing of effective art therapies in forensic setups.

In another review, researchers studied neuropsychological testing in forensic setup. Memory, cognitive level tests, and symptom validity were the most commonly assessed constructs in the reviewed studies (da Silva & Hamdan, 2022). The review indicates that studies have a higher number of male participants and a limited number of neuropsychological constructs. This opens another avenue for forensic neuropsychological research. To have a wholesome understanding of the effects of being in prisons or related setups and being neuropsychologically or psychiatrically impaired in such setups, a wider range of populations should be included in future studies, such as females, transgenders, and inmates from different age groups. Neuropsychological studies can help us paint a better picture of the effects of being involved in forensic setups like in courtrooms, mental health care facilities in prison, or being in the management part of the prison and also those who are related to those directly involved.

Despite its ardent need, surveys have indicated that there is lesser use of neuropsychological testing and assessments by attorneys in legal setups. Studies have shown malpractices by legal practitioners and also sometimes by forensic neuropsychologists in order to manipulate the case in favor of their client by sometimes showing malingering or covering up the symptoms(Essig et al., 2001). Extensive research in the field is required to develop good ethical policies for the deployment of appropriate techniques and assessment methods in practice in both clinical and legal setups.

The novel tools and data acquisition methods are rapidly expanding, which makes it important for forensic neuropsychologists to keep themselves up-to-date with advancing technologies to have validated evidence. The evolution of digital technologies for the assessment in neuropsychology is going to be another huge area for research and implementation.

References

Abbing, A., Haeyen, S., Nyapati, S., Verboon, P., & van Hooren, S. (2023). Effectiveness and mechanisms of the arts therapies in forensic care. A systematic review, narrative synthesis, and meta analysis. *Frontiers in Psychiatry, 14*, 1128252. https://doi.org/10.3389/fpsyt.2023.1128252

Albright, T. D. (2017). Why eyewitnesses fail. *Proceedings of the National Academy of Sciences of the United States of America, 114*(30), 7758–7764. https://doi.org/10.1073/pnas.1706891114

Arias-Carrión, O., Stamelou, M., Murillo-Rodríguez, E., Menéndez-González, M., & Pöppel, E. (2010). Dopaminergic reward system: A short integrative review. *International Archives of Medicine, 3*, 24. https://doi.org/10.1186/1755-7682-3-24

Blonigen, D. M., Carlson, S. R., Krueger, R. F., & Patrick, C. J. (2003). A twin study of self-reported psychopathic personality traits. *Personality and Individual Differences, 35*(1), 179–197. https://doi.org/10.1016/S0191-8869(02)00184-8

Chow, T. W. (2000). Personality in frontal lobe disorders. *Current Psychiatry Reports, 2*(5), 446–451. https://doi.org/10.1007/s11920-000-0031-5

Comaty, J. E., Stasio, M., & Advokat, C. (2001). Analysis of outcome variables of a token economy system in a state psychiatric hospital: A program evaluation. *Research in Developmental Disabilities, 22*(3), 233–253. https://doi.org/10.1016/S0891-4222(01)00070-1

Cornish, D. B., & Clarke, R. V. (1987). Understanding crime displacement: An application of rational choice theory. *Criminology, 25*(4), 933–948.

Crockett, M. J., Clark, L., Tabibnia, G., Lieberman, M. D., & Robbins, T. W. (2008). Serotonin modulates behavioral reactions to unfairness. *Science (New York, N.Y.), 320*(5884), 1739. https://doi.org/10.1126/science.1155577

Dabbs, J. M., & Morris, R. (1990). Testosterone, social class, and antisocial behavior in a sample of 4,462 men. *Psychological Science, 1*(3), 209–211. https://doi.org/10.1111/j.1467-9280.1990.tb00200.x

Davies, G. M., & Beech, A. R. (2017). *Forensic psychology: Crime, justice, law, interventions.* John Wiley & Sons.

De Boer, S., Caramaschi, D., Natarajan, D., & Koolhaas, J. (2009). The vicious cycle towards violence: Focus on the negative feedback mechanisms of brain serotonin neurotransmission. *Frontiers in Behavioral Neuroscience, 3*, 977.

Essig, S. M., Mittenberg, W., Petersen, R. S., Strauman, S., & Cooper, J. T. (2001). Practices in forensic neuropsychology: Perspectives of neuropsychologists and trial attorneys. *Archives of Clinical Neuropsychology, 16*(3), 271–291. https://doi.org/10.1093/arclin/16.3.271

Fielenbach, S., Donkers, F. C., Spreen, M., & Bogaerts, S. (2018a). Effects of a theta/sensorimotor rhythm neurofeedback training protocol on measures of impulsivity, drug craving, and substance abuse in forensic psychiatric patients with substance abuse: Randomized controlled trial. *JMIR Mental Health, 5*(4), e10845.

Fielenbach, S., Donkers, F. C., Spreen, M., Visser, H. A., & Bogaerts, S. (2018b). Neurofeedback training for psychiatric disorders associated with criminal offending: A review. *Frontiers in Psychiatry, 8*, 313.

Gregory, S., Simmons, A., Kumari, V., Howard, M., Hodgins, S., & Blackwood, N. (2012). The antisocial brain: Psychopathy matters: A structural MRI investigation of antisocial male violent offenders. *Archives of General Psychiatry, 69*(9), 962–972.

Gudjonsson, G. H. (2003). Psychology brings justice: The science of forensic psychology. *Criminal Behaviour and Mental Health, 13*(3), 159–167.

Gudjonsson, G. H., & MacKeith, J. A. (1988). Retracted confessions: Legal, psychological and psychiatric aspects. *Medicine, Science and the Law, 28*(3), 187–194.

Konicar, L., Veit, R., Eisenbarth, H., Barth, B., Tonin, P., Strehl, U., & Birbaumer, N. (2015). Brain self-regulation in criminal psychopaths. *Scientific Reports, 5*(1), 9426.

LePage, J. P., DelBen, K., Pollard, S., McGhee, M., VanHorn, L., Murphy, J., Lewis, P., Aboraya, A., & Mogge, N. (2003). Reducing assaults on an acute psychiatric unit using a token economy: A 2-year follow-up. *Behavioral Interventions: Theory & Practice in Residential & Community-Based Clinical Programs, 18*(3), 179–190.

Mazur, A., & Booth, A. (1998). Testosterone and dominance in men. *Behavioral and Brain Sciences, 21*(3), 353–363.

Mazur, A., & Booth, A. (1999). The biosociology of testosterone in men. *Social Perspectives on Emotion, 5*, 311–338.

Moffitt, T., Caspi, A., Fawcett, P., Brammer, G. L., Raleigh, M., Yuwiler, A., & Silva, P. (1997). Whole blood serotonin and family background relate to male violence. In A. Raine, P. A. Brennan, D. P. Farrington, & S. A. Mednick (Eds.), *Biosocial bases of violence* (pp. 231–249). Springer US. https://doi.org/10.1007/978-1-4757-4648-8_14

Narvaes, R., & Martins de Almeida, R. M. (2014). Aggressive behavior and three neurotransmitters: Dopamine, GABA, and serotonin—A review of the last 10 years. *Psychology & Neuroscience, 7*(4), 601.

Oehler, S., Berman, A., Gay, C., Manguso, R., & Espinoza, J. (2018). An analysis of the correlates of aggression in a social learning program for severely and persistently mentally ill inpatients. *Archives of Psychiatric Nursing, 32*(1), 39–43.

Outsem, R. (2014). Novel Neuropsychological Opportunities in the Treatment of Antisocial Behavior. In J. H. Gallo (Ed.), *Antisocial Behavior: Etiology, Genetic and Environmental Influences and Clinical Management* (pp. 15–42).

Pope, H. G., Kouri, E. M., & Hudson, J. I. (2000). Effects of supraphysiologic doses of testosterone on mood and aggression in normal men: A randomized controlled trial. *Archives of General Psychiatry, 57*(2), 133–140.

Raine, A. (2002a). Annotation: The role of prefrontal deficits, low autonomic arousal, and early health factors in the development of antisocial and aggressive behavior in children. *Journal of Child Psychology and Psychiatry, 43*(4), 417–434.

Raine, A. (2002b). Biosocial studies of antisocial and violent behavior in children and adults: A review. *Journal of Abnormal Child Psychology, 30*, 311–326.

Silva, L. V. da, & Hamdan, A. C. (2022). Neuropsychological assessment in the forensic context: A scoping review. *Brazilian Journal of Forensic Sciences, Medical Law and Bioethics, 12*(1), Article 1. https://doi.org/10.17063/bjfs12(1)y202253-74

Stevens, A. (2010). Introducing forensic democratic therapeutic communities. In R. Shuker & E. Sullivan (Eds.), *Grendon and the emergence of forensic therapeutic communities: Developments in research and practice* (pp. 7–24). Wiley-Blackwell.

Sweet, J. J., Boone, K. B., Denney, R. L., Hebben, N., Marcopulos, B. A., Morgan, J. E., Nelson, N. W., & Westerveld, M. (2023). Forensic neuropsychology: History and current status. *The Clinical Neuropsychologist, 37*(3), 459–474. https://doi.org/10.1080/13854046.2022.2078740

Tolisano, P., Sondik, T. M., & Dike, C. C. (2017). A positive behavioral approach for aggression in forensic psychiatric settings. *The Journal of the American Academy of Psychiatry and the Law, 45*(1), 31–39.

Umbach, R., Berryessa, C. M., & Raine, A. (2015). Brain imaging research on psychopathy: Implications for punishment, prediction, and treatment in youth and adults. *Journal of Criminal Justice, 43*(4), 295–306.

Veenema, A. H., & Neumann, I. D. (2007). Neurobiological mechanisms of aggression and stress coping: A comparative study in mouse and rat selection lines. *Brain Behavior and Evolution, 70*(4), 274–285.

Wortley, R., & Mazerolle, L. (2009). Environmental criminology and crime analysis: Situating the theory, analytic approach and application. *Crime Prevention and Community Safety: An International Journal, 11*. https://doi.org/10.1057/cpcs.2008.22

Part 2

6 Memory and Neuropsychology

Introduction

We are the sum of all our experiences. Everything that we have lived through forms the basis for how we will behave in the future as well. Therefore, think of how this record functions in our brains. It is a sort of time travel, and our brain makes sure that we are stuck in time, almost replaying the same instances over and over. The alternative is just losing that record altogether. Disorders like amnesia and Alzheimer's dementia mean that these record dysfunctions are leading to general disarray. When we refer to memory, there is no extensive system but a series of subsystems that correspond to the various processes that are involved in the process of memory. Memory is the process of encoding, storing, and retrieving information. Our discussion in this chapter will center around how the brain stores and restores this information when necessary. We will also cover the processes of memory and cover some of the disorders that are predominant in this function. These include disorders like amnesia, along with neurodegenerative disorders like Alzheimer's. Various methods of rehabilitation will be debated at the end of the chapter.

The influential idea, in the beginning, was that there was a localized representation of memory in the brain that led to actual physical changes in the brain itself. These connections mean that there are connections that grow and form between the areas of the brain. Ivan Pavlov, the Russian physiologist, pioneered the research on classical conditioning. The original experiment was one on salivation and the digestive glands. It was the experiments on the gastric glands that led to the observation that the dogs would salivate before he could give them food. He was intrigued by this response and designed his later experiments to study the same. He believed that even before they saw the food, they would still recognize that he was coming with the food by the sound of his footsteps. He termed the food as the unconditioned stimulus (UCS) and the salivation as the unconditioned response (UCR). The UCS was paired with the bell, which was the conditioned stimulus (CS). When the UCS and the CS are paired together, the dog learns a conditioned response (CR) at the sight of the CS.

DOI: 10.4324/9781032640839-8

Another form of learning is operant learning, where the responses are followed by reinforcement or punishment that serves to either strengthen or weaken a behavior. Reinforcers are events that increase the chances of the response happening again, while punishment is those events that reduce the likelihood of the response occurring again. A relatable example would be if you study for an exam and you do well in it. Those marks would become a reinforcer and increase the chances of you studying for exams in the future. Nevertheless, if, despite your best efforts, you do not get the desired result, those same marks would become a punishment. It would reduce the chances of you preparing well for future exams. The main difference between these two types of conditioning is that one cannot learn new behavior through classical conditioning but that can happen through operant conditioning. The chances of the behavior happening or not happening depends on the reinforcer or the punishment. However, in the classical conditioning scenario, it would happen whenever the CS is presented, given that it has not been a lot of time since they were last paired.

The use of classical conditioning also showed many other properties of learning and memories. There was progress in how quickly the CR could be elicited over time. Another observation was that if this stimulus-response relationship was not elicited regularly, there was a chance that it could go extinct. These led Pavlov to believe that conditioning strengthened the area that was related to the CS and the UCS centers in the brain. In order to prove this hypothesis, Karl Lashley tried to find engrams or physical representations of this learning in the brain. He devised an experiment using rats in a maze and tested them on a brightness discrimination task. When sufficient learning had occurred, he proceeded to cut various areas on the rats' brains and had them rerun the maze. It was found that there was no loss in learning, and he could not narrow it down to any single area in the cortex. This research led to the principle of equipotentiality, which stated that all parts of the cortex function equally within the cortex for complicated tasks (Lashley, 1950). The critical observation here is that the tasks that Lashley chose to find the engram were difficult and complex. These would involve other areas and require cooperation between the various regions of the brain. Lashley's conclusion was based on the assumption that learning could only occur in the cortex and that these are all mostly the same. His contemporaries, which came after him, reached different conclusions once these were discarded (Kalat, 2015).

Richard Thompson gave a much easier experiment that involved the use of eyelid responses in rabbits. These were also much simpler learning which was a reflex located in the cerebellum instead of the cerebrum. They first played a tone and then blew a puff of air into the rabbit's eye to evoke the blinking response. They also recorded the brain in order to understand the areas that change during learning. If any of these areas are blocked or destroyed, then in the sequence of areas, there would not be any further learning in that area or the areas that are connected after it. Thompson identified a nucleus known as the lateral interposed nucleus as being essential for learning. When this area was suppressed, the rabbits did not show any response. However, this was the area that was connected to that particular reflex, along with other regions of the cerebellum, like the red nucleus that receives its inputs from the cerebellum. When this area was suppressed, there was no response, but when it got reactivated, there was a robust learned response. Even the hippocampus was identified as a seat of learning in the brain (Thompson, 1986). The search for the engram is still on, and some have been found, but there are still others that remain elusive. There has been some progress concerning engrams for habit learning, and emotional learning has been found. These required the use of molecular biology and different recording approaches. Lashley was not too far off when he had declared an engram several decades earlier (Eichenbaum, 2016).

Types of Memory

For a long time, there were no clear distinctions between learning and memory. It was Donald Hebb (1949) who theorized that no one mechanism could account for all the variations that were possible in learning. The same mechanism that made sure that a person could repeat a series of numbers given to them could also remember and recall things that had happened a long time ago. He was the first to attempt to differentiate memory in terms of short-term and long-term memory. Short-term memory is for any events that occurred immediately, whereas long-term memory would cover events that might have happened a long time ago. This difference is also showcased in the apparent capacity of the two types of memory. Short-term memory has a limited capacity, and one cannot recall more than seven items at a time. The capacity of long-term memory is seemingly infinite, and there is no right way of estimating it yet. Everything in the short term is subject to rehearsal, while the long term is based on cues. Also, based on the limited capacity, anything which gets interrupted in the short-term memory gets lost forever, whereas the long-term memory does not permanently lose anything. There are more chances of recovery in this case.

This distinction means that whatever comes up in short-term memory must be converted to long-term memory. However, there is a process of consolidation that not every piece of information goes through. There are also instances of information that do not fall cleanly into either of these types.

In order to resolve the problems that are presented by the conceptualization of short-term memory, Baddeley and Hitch proposed the working memory model as a means of storage of memory that occurs while working. It refers to the system necessary to keep things in mind while performing complex tasks. The rise of the working memory model was derived from the information-processing models and the concurrent rise of technology worldwide. The technology is the most appropriate parallel for the working memory systems still is the computer. Its evolution from short-term memory to a working memory system also saw the inclusion of cognitive neuropsychology, which provided empirical proof to the theories. In particular, was the case of H.M. (Henry Gustav Molaison), whose case was the basis for our understanding of the different types of memory and the conceptualization of amnesia. The working memory is composed of three parts: the visuospatial sketchpad, the phonological loop, and the central executive, – the central executive functions as the supervisory system for the phonological loop and the visuospatial sketchpad. The visuospatial sketchpad is a conceptualization of memory that depends on the sense of vision and how the information tends to last visually for a specified period. The phonological loop, similarly, is based on the rehearsal as a means of repeated vocalizations of the information. It tends to store all kinds of auditory input, while the visuospatial sketchpad stores all the visual input. Attention is a relatively scarce resource that cannot be distributed equally for all tasks. It falls to the central executive to direct this attention to either of these two systems.

A modern example of this would be a one-time password necessary for any online transaction. By the time we get the message and then later enter it into the corresponding textbox, we tend to repeat the sequence of digits repeatedly. This repetition keeps the information in the phonological loop and retains it for that time as necessary. We do not remember this later as it becomes irrelevant. The delayed response task is a test of working memory that requires responding to a stimulus that one heard or saw a while earlier. An increased level of activity was observed in the prefrontal cortex during the delay, which indicates that the information was being stored. The stronger this activation of the prefrontal cortex, the better their performance in the task (Baddeley, 2010). As people continue to age, there is an overall decline in many parts of their functionality. Concerning the functioning of working memory in older adults, Gazzaley et al. (2005) found a possible cognitive reason explaining its decline. Working memory and attention are two processes that are interconnected and need each other to function optimally. A recently developed task showed two scenes and two faces, and in its three conditions, the participants had to either concentrate on the faces or the scenes. A third condition was a passive view of these stimuli. Depending on the instructions given for each of these scenes, the activity level would change. When the younger participants were given instructions to remember the stimuli, they showed high activity in the prefrontal cortex. This activity level was higher than when they were instructed to ignore certain stimuli. The older participants, in comparison, showed no difference in enhancing activity to remember the stimuli but had a decline when it came to suppressing stimuli. The older participants are not able to restrict task-irrelevant information. This difference was demonstrated by the blood oxygen level-dependent (BOLD) signal. Since attention is, therefore, seemingly going to both remembering and forgetting at the same time, the function of the working memory is affected. This level of declining activity adversely affects brain function but is compensated with more activity in other areas. Another factor that usually impairs working memory is traumatic brain injury (TBI).

A modified version of the paced auditory serial addition task was used where the participants were given a series of numbers at the rate of one per two seconds. The first two numbers were added and then this went on until the sum reached ten. The sums were not done out loud, and whenever the participant reached this sum, they would indicate this by raising their index finger. It is a task that demands a significant load on the memory reserves. A separate control task was also used on healthy controls and patients with TBI. On a behavioral level, the patients with TBI made significantly more errors than the controls in the past. However, this was on a comparable level, indicating that there was some function of working memory left. The functional magnetic resonance imaging (fMRI) data showed that patients with TBI demonstrated altered cerebral activation during the tasks when compared with the healthy controls. The activation was lateralized to the right in the patients with TBI as opposed to the left in controls (Christodoulou et al., 2001).

Apart from working memory, it is now necessary to explore the other types of memory as well. This distinction is purely to enhance our understanding of the various types of information that the brain is capable of storing. The human memory has a sensory memory stored as well that corresponds to the sensory system. The eyes store a memory of the image for less than a second before this is forwarded to the brain for processing. Other sensory organs also work similarly. Long-term memory is a massive storehouse of information, and cataloging its inventory would be beneficial for research. There are explicit and implicit memory stores. The former is a more conscious form of memory and deals with information that underwent consolidation to long-term memory. The latter, which is implicit memory, is the type of memory that does not necessarily follow the flow of encoding, storing, and retrieval. Many tasks or skills that usually follow a step-by-step procedure are usually in this area.

An example of this would be your actions when involved in complicated procedures like driving a car. When we speak of explicit memory, there is declarative memory. It deals with the memory of facts and events, along with memories of those involved in them. It is further divided into episodic memory and semantic memory. Semantic memory is involved in facts and concepts, and these would constitute our knowledge or learning and is essential for our language skills.

Episodic memory is about events and experiences that we might read about in the newspapers or have personally experienced and lived through. These can be recalled and, in essence, be re-experienced mentally. Tulving (1972) envisioned the two types as being different ways of processing information. These select the type of information from the senses and retain different aspects of the information. He posited that episodic memories were linked to temporal aspects of information. This chronology also made it susceptible to change as every access to this store of memory would also constitute an input. The episodic memory is an autobiography that is being updated for as long as a person is alive. Semantic memory is the encyclopedia of everything that is around us in the world. Some of the neural components that are involved in the functioning of episodic memory are the cortex near the hippocampus, which is the perirhinal cortex (PRC), parahippocampal cortex (PHC) and the entorhinal cortex, other cortical structures, and the hippocampus. The entorhinal cortex is the primary contact between the hippocampus and the neocortex. It has multiple layers connecting the hippocampus to the surrounding cortical areas. These receive processed inputs from the various sensory systems and any cognitive processes. The deeper layers receive the output from the hippocampus, which is then communicated to the other areas. In these areas, the PRC plays a role in visual object recognition, while the PHC takes care of how the local environment is perceived and processes information in this area. Lastly, the hippocampus forms new memories based on the inputs it receives from these areas and then retrieves them as well (Camina & Güell, 2017). The neurobiology of long-term memory is similar to episodic memory, but semantic memory is different, and this shall be explored in the coming paragraphs.

Semantic memory might be a vast store of knowledge and information. However, it directly contributes to everything that we as humans do in our daily lives. The use of this conceptual knowledge to specialize and apply to problems in everyday life is a human trait. The neurobiology of semantic memory is based on various neuroimaging studies that show that there are sensory, emotional, and motor systems in language comprehension that is specific to modalities. Along with this are regions of the inferior parietal lobe and much of the temporal lobe. These areas also lie in the convergence of multiple perceptual processing streams, which indicates that there is a dedicated neuroanatomical model for processing semantic information (Binder & Desai, 2011). These process information from various systems, and its integration leads to the creation of the knowledge store. The last type of memory is a relatively new and novel approach. It deals with events or information that has yet to occur or will occur in the future. This is called prospective memory. According to Kvavilashvili and Ellis (1996), it has been defined as "remembering to do something at a particular moment in the future or as the timely execution of a previously formed intention" (p. 25). These are based on the intentions that are formed in anticipation of a particular time or the chance of an event occurring. Intentionality forms the memory of acting. There needs to be a particular duration between the formation of the intention to act and the action itself. This is the time necessary to form the memory. In order to understand the neurobiology of prospective memory, a study was done using older adults divided into composite scores for the frontal lobe and the temporal lobe. These were based on tests like the Wechsler Memory Scale (WMS) and Paired Associate Tests. The frontal lobe was tested explicitly based on a modified Wisconsin Card Sorting Test and Controlled Oral Word, along with Wechsler's Adult Intelligence Scale. The results showed that older adults who had lower scores on the tests associated with the frontal lobe had diminished capacity in terms of prospective memory. This means that the frontal lobe functions are essential to sustain prospective memory (McDaniel et al., 1999).

Consolidation and Reconsolidation of Memory

Memory consolidation has been associated with many necessary factors. One such model of consolidation meant that there was a reason why the consolidation process was slow. This slow process has an adaptive function in order to modulate the consolidation process that buffers for the effects of stress or emotions. The hippocampus in the medial temporal lobe (MTL) is vital to memory consolidation as the amygdala is another structure that is also responsible for memory consolidation. It is the emotional part of the brain and is the main reason that memory with strong emotional counties is more likely to be remembered for a longer duration as well. Other hormones like cortisol and epinephrine are essential for memory consolidation as they interact with the hippocampus and the amygdala (McGaugh, 2000). The MTL houses the hippocampus and the associated areas like the entorhinal cortex, PHC, and the PRC. It was also taken to be the area that memory was stored in.

Nevertheless, it has been increasingly recognized as the seat of consolidation rather than storage. The entire MTL system would technically function as a temporary store of memory, and the neocortex is the permanent area for long-term memory. More connections between the MTL and the other areas and also within the MTL would mean that the consolidation process is fast (Alvarez & Squire, 1994).

Assessment of Memory

Theories

Miller's Magical Number. This theory is one of the key components in the history of memory and, more specifically, the limits that this has on itself. These capacities are on the various systems of memory. He gave a talk that set a limit on immediate memory at seven, give or take two chunks. Chunks were the measure that Miller (1956) set on the information pieces that the immediate memory would hold. These chunks are evident when we think of the way we memorize information in our everyday lives. Phone numbers are not read out in the single string of numbers that they essentially are but in the form of two or three chunks. 98######56, is usually read as 98## - ## - ##56.

Memory Decay. When Miller gave the magic number, there were speculations about how long information could be retained if the capacity was limited. Trigrams were used, which are lists of meaningless three letters, which made it difficult for the participants to assign meanings. This interference prevented a deeper level of encoding and the addition of an interference task, where the participants were also asked to count down from a number, made rehearsal difficult. This study showed that while the capacity is limited within the immediate short-term memory, it also loses information quickly unless supported with sufficient rehearsals (Peterson & Peterson, 1959).

Interference Theory. This hypothesizes that memory is more likely to decay by being forgotten due to interference with other types of memory. There can be two types of interference. When new information interferes with older information, it is known as retroactive interference. Meanwhile, when information that has already been stored within the brain interferes with the acquisition of new information, it is called proactive interference. Underwood and Postman (1960) demonstrated this effect when they gave a group of participants with a list of words with stimulus and response words. The instructions were to give the response words when presented with their corresponding stimulus word. A second group in the study was also given a similar list, but two were also presented with a second list of word pairs that had to be memorized as well. As expected, the group that learned a second list of words was less accurate than the group that learned just the first list. It also makes a case for the limited reserves of attention and how this would hamper recall as well.

Types of Tests

Following are some of the commonly used tests for General Memory:

Wechsler Memory Scale

The Wechsler Memory Scale is a test in neuropsychology that assesses the verbal and nonverbal components in adults. It was first developed in 1917. It assesses learning, memory, and working memory, with the latest revision in 2009. It was updated in the last decade, indicating that it continues to be a preferred tool for clinicians. It assesses visual, verbal, and auditory memory along with other modalities as well. This test is appropriate for ages 16–89 and takes about an hour to administer. Although there have been reports that the delayed recall index is still not measured in the scale, the WMS is not the only scale to measure from this deficiency. While the validity of the delayed recall index has come into question, the test still enjoys a robust internal consistency score ranging from 0.70 to 0.90 (Kent, 2013).

Rivermead Behavioural Memory Test

This test was developed by Wilson et al. (1997). It comprises many subtests that provide objective measures of everyday performances. It has excellent psychometric properties and has about 12 subtests that range from verbal, visual, and visuospatial memory in immediate, delayed, and prospective conditions. These include but are not limited to names, faces, orientation, dates, and pictures (Moradi et al., 1999).

Memory Assessment Scales

Developed as an alternative to the popular WMS, the Memory Assessment Scales is a comprehensive battery for memory functioning in adults. It comprises several subtests that range from list learning, verbal span, and visual reproduction to names, faces, and visual recognition. The test is also conducted in the delayed condition as well. Norms are available for ages 18–90 (Williams, 1991).

Test of Memory and Learning

Developed by Reynolds and Bigler (1994), this test consists of eight core subtests and boasts of a shorter administration time than other tests in the field. It gives three indices across verbal, non-verbal, and composite memory. In addition to the core tests, there are also additional indexes that measure attention and concentration, learning, and free recall. Another unique property of this test is that it can be administered to ages 5–59.

Implicit Memory

We have discussed in this chapter, that all forms of memory can be neatly divided into either implicit or explicit forms of memory. Most of the research done in this field tends to focus on explicit memory. It is due primarily to the fact that this is much more observable and reliable than implicit memory. However, there are also the following tests that measure implicit memory.

The implicit **Association Test** is a measure of implicit attributes that might be associated with two target concepts. It involves choosing between two possible choices and following measures of the kind of attribute that we might hold with regard to the two concepts. It is an experimental test that has its roots in the subfield of social psychology. Since this is an implicit measure, it requires inputs in a small duration. These tests are usually linked with sociodemographic details and seek to uncover the implicit associations between them (Greenwald et al., 1998).

Word Stem Completion is another class of implicit memory measures. In terms of research, this tends to measure whether there is any carryover effect between tests that are unrelated to each other. It also helps us understand the link between explicit and implicit forms of memory and retention. The word stem would include an incomplete problem that would have to be answered or solved based on a question or a cue based on previously presented information. This information could either be connected to the task at hand or be in the realm of general knowledge as well. Priming is usually not taken as a part of the tests (Roediger et al., 1992).

Delayed Matching Sample Task 48

It is a visual recognition memory test for memory changes that typically accompanies Alzheimer's disorder. This test is easy to administer and enjoys excellent reliability and validity with the Wechsler Story Recakk Task. There are a series of 48 images colored and computer generated that are presented to the participant in sequential order. Later, these are divided into three sets consisting of 16 pairs. Each of these sets is either of three categories. The first one is unique, where the target image is presented with a distractor image that has no semantic or lexical connection to it. The second set is the matched set, where the target image and the distractor image look similar in terms of color and images. The last set is abstract that has no basic image; a typical example of this set would be the mandala. None of the distractor images is repeated through the three sets. The correct response is given in terms of percentages, where 100% means that all 48 of the target images have been identified correctly. In contrast, 50% may be achievable if the participants answer at random or by chance (Rullier et al., 2014). Episodic memory is the most affected in patients with Alzheimer's disorder, and this test provides an ideal diagnostic tool that is easy to administer as well.

Grober & Buschke Test

The Grober & Buschke Test is a measure for screening dementia. It consists of enhanced cued recall along with tests for naming and spatial location. Items used are line drawings that can be easily cognized. Dominant items in these categories were avoided to reduce guessing. Sixteen items were presented four at a time, with an item in each quadrant of the cards. The participants are asked to point at items on the cards when the category is verbally given. Once all the items on the card have been identified, they are asked to recall the items based on the categories as cues. The fact that the participant can search items successfully indicates that deep semantic processing was possible, while successful completion of the immediate recall indicates encoding and retrieval (Ros et al., 2018).

Autobiographical Memory Test

Autobiographical memory test measures a particular component of autobiographical memory called overgeneral autobiographical memory (OGM). The OGM refers to the difficulty in retrieving specific memories, which is instead replaced with an overly general memory. The OGM has also been linked with psychopathologies of affective disorders, schizophrenia, and personality disorders. The participant in this test is presented with a series of cue words and then asked to give a specific memory or event from their life for each. The procedures for the administration of the tendency vary across studies depending on the number of cues used, their emotional valence, the presence of prompting for a specific memory, and the inclusion of a time limit. Typically, 10–20 cues are given in studies, while most of the words are either positive or negative. Neutral words are avoided in these studies. Some examples of the words that are used in these studies are happiness, friendship, failure, and worry. These tend to be divided along with either the positive or negative words. These are usually done in conjugation with events that might have happened a week prior (Ros et al., 2018).

Cambridge Prospective Memory Test

Assessment of prospective memory (PM) depends on aspects like intention formation, storage, and timely retrieval. The first is difficult to assess, but the second has many tests to its credit. Timely retrieval would require tasks of sustained attention that are necessary to retrieve. This test is more suitable for assessment in real-life scenarios. Apart from individually assessing these processes that form a part of the prospective memory, some tools cater to it as a whole. The Rivermead Behavioural Memory Test has PM items like remembering to deliver a message and appointment dates in its items. The Cambridge Prospective Memory Test has three time-based and three event-based tasks that must be acted upon within 30 minutes. However, this does not demonstrate adequate reliability. It has prompted research into naturalistic tasks for prospective memory, like making phone calls five times daily at their designated times (Fish et al., 2010).

Rey Auditory Verbal Learning Test

It was developed in the 1940s and currently consists of a list with 25 words that are nouns, one per second. The person being tested has to recall all of these words in any order. The entire procedure is then carried out four more times. Finally, the examiner presents a second list of 15 words. After this, the participant has to recall the words from the first list that was presented. The Rey Auditory Verbal Learning Test (RAVLT) is a relatively simple test to administer and is useful in assessing verbal learning, memory, and proactive and retroactive inhibition as well.

Benton Visual Retention Test

The Benton Visual Retention Test by A. L. Benton was developed in 1963. It consists of ten cards, each of these has a few geometric designs on them. The cards are only exposed for a duration of 10 seconds. The subject must draw the figure immediately after the card has been removed. The test requires spatial conceptualization, immediate recall, and visuomotor reproduction. The test is unique in that it can also estimate preservative errors (Steffens et al., 2003).

Rehabilitation of Memory

Memory rehabilitation emphasizes on lessening the impact of memory impairment and improving activities of daily living. Usually, after brain injury, there is some resultant dysfunction along with spontaneous recovery as well. It may take several weeks to months in most cases, but the process occurs in all cases.

Memory and Brain Plasticity

Brain plasticity has provided hope that despite the destruction of neurons, there is still enough functionality left that makes sure that the other parts of the brain can take over functionality. It is also known as neural plasticity, which is the ability of the nervous system to alter its activity in response to stimuli. This process also reorganizes the structure, functionality, and connections between synapses. The modification usually occurs in terms of the strength of transmissions. It also plays a vital role in learning and during traumatic brain injury (Mateos-Aparicio & Rodríguez-Moreno, 2019). The activity defines the level of morphological modification, which reflects the "Use it or lose it" phenomenon. In total, there are four ways that neural plasticity may occur in conjugation with new memories. Most neural circuits may be parts of the brain that appear relatively stable to these effects. However, the neural circuitry in the cortical regions associated with learning and memory seems to undergo a process of dynamic modification based on learning and memory. It is either a result of synaptic strengthening, synaptic weakening, synapse formation, remodeling, and neurogenesis. Synaptic strengthening, in particular, seems to result from long-term potentiation. The synaptic structure formation unique to the long-term potentiation (LTP) is vital in memory consolidation. Similarly, the effects of synaptic weakening indicated that through the long-term depression of activity, there would be functional elimination of areas that are not necessary for memory consolidation. It means that those forms of memory that were not accessed as frequently could have been shifted into photos of those that were accessed regularly. This also fine-tunes the neural mechanisms that are associated with the other processes of neural plasticity. All four of these mechanisms do not work independently of each other. An example is that any area undergoing the effects of LTP would witness the neighboring regions have long term depression (LTD). Another necessary clarification in this process is that the neural plasticity process is possible only in cortical regions that are attuned to learning and memory. Failure of these mechanisms with age inevitably leads to memory deficits that are associated with healthy aging. It also opens up the brain to pathological disorders like neurodegenerative disorders, neural disorders, and psychiatric disorders (Bruel-Jungerman et al., 2007).

Restorative Approach

One form of rehabilitation depends on cognitive retraining. This retraining involves the use of repeated exercise for tasks that challenges the cognitive function that is being trained. It shows that retraining memory as a cognitive function also involves the strategies and skills based on mnemonic devices, repetitive practice, or using external devices.

Compensatory Approach

An approach that looks to modify the environment and other areas rather than the person with a memory deficit is the compensatory approach. Four primary techniques used in this approach are enhanced learning, mnemonic strategies, external aids, and environmental modification.

Mnemonic strategies are essential techniques for memorizing large amounts of information. A useful method for this is the first letter technique, where the first letter of the information leads to a particular word. It also acts as a cue for the same. A famous example of this would be the VIBGYOR, where each of the letters in the word is the first letter of the colors of the rainbow. Another commonly taught method is the method of the loci, where people with memory impairments are taught to associate a familiar place or a location with memory information.

The use of external aids can also be an alternative that many people with memory impairment seem to adapt on their own. It has its shortcoming in that learning to use external aid might become too cumbersome, especially with mobile phones. There have been mixed results when it comes to memory. While there is no argument against phones for people with memory impairment, its effects on memory itself are questionable. It can be observed that people have started depending more on their phones rather than their brains for storing information. Birthdays to phone numbers, all of which were usually remembered, have been lost over the years. It is a miracle if one can remember their phone numbers now! Caution must be exercised so that the system or aids that are used are just that. If the aid becomes a replacement for the system that it aimed to aid, it would not be just an aid.

Environmental modifications is the final method that is used when the person in question is not able to use any other methods. This technique is particularly needed when the person with memory impairment is unable to match the other methods. These would mean the use of light posts or color-coded signs for bathrooms. Providing a highly structured environment that is arranged around a daily routine also helps. Repetition of information is an excellent form of learning, and this often leads to even the most severe cases of memory deficit in learning routines and other daily tasks.

Rewriting Traumatic Memory

Memory is pretty malleable, and every time this is accessed, the brain itself tends to rewrite and alter it all the time. This malleability is primarily in terms of neuroplasticity, which allows this to be possible. This alteration is difficult for memories that specifically relate to fear. The study is sought to understand if fear could be attenuated by rewriting the memory of fear to one of safety. Chemicals, genes, and a specific engram might hold the answer for the same. The important thing is to sever the connection between the fear and the memory. The areas under question were the dentate gyrus which plays a vital role in encoding, recalling, and attenuation of context-based fear and its retrieval. Memory recall, in particular, activates a subset of neurons in the dentate gyrus. Continued activation of these neurons ensures that fear attenuation is an ongoing process. This also implicates inhibitory memory traces that exist within the original memory trace, as suggested by Pavlov. It would seem that reconsolidation of memory would be easier than extinction learning (Khalaf et al., 2018).

Research Perspective

In the above sections, we understood the neuropsychology of memory, what comes under when we say the word "memory," its types, processes, and assessment methods. Early understanding of brain parts underlying memory function indicated different subsystems responsible for different memory functions. The scientific community has been constantly working on designing new experiments and refining the assessment methods to better our understanding of how different memories are processed, from encoding to retrieval within the brain. Studies in recent decades have shown that brain plasticity impacts the functional organization of the brain, and when that happens, memories are indeed subject to change. It has been understood that not individual subregions but a network of complex connections and layers of organizations within the subregions are how memories are processed. Research is now focusing on understanding these processes that, in turn, shape the behavior driven by memories. Over the last few decades significant advancements in neuroimaging techniques have allowed researchers to gain valuable insights into memory processes and also in developing efficient treatment plans in case of memory dysfunction. There are a variety of experiments that have been designed to study different aspects of memories while using various methods available for neuropsychological assessment.

Advancements in assessment methods have given researchers the opportunity to explore different aspects of core memory processes. Some of the research areas include suppression of moral, immoral autobiographical memories, computational models of semantic memories, encoding and retrieval of emotional memory, consolidation of memory during sleep, recognition memory, visual working memory, and memory dysfunction in different sets of populations. The fMRI studies have delved into exploring the association of thalamic structure with the MTL. An fMRI study using the recognition memory paradigm investigated familiarity and recollection of objects, faces, and scenes. The results showed activity in thalamic regions, including mediodorsal thalamus, anterior thalamus, pulvinar thalamic nuclei, and ventral posteromedial thalamic nuclei, along with activation in various parts of the MTL (Kafkas et al., 2020).

Further studies will be needed using assessment methods with a higher spatial resolution to study these effects in further detail within thalamic nuclei. Such studies give us a broader understanding of brain regions working together to support various memory systems, which helps us understand the effects of memory disorders better. Furthermore, another facet of this research category entails the investigation of fluid and flexible semantic memory sensitive to different aspects of information available. Since concept formation and deriving meaning from it is an integral part of cognition, it becomes evident to assess the current status of semantic models and find the gaps for further research (Kumar, 2021). Their review article on traditional and computational models of semantic memory has identified the need to explore further the role of communication, social cognition in developing semantic knowledge, age-related challenges in semantic tasks, and the development of multilingual semantic models. Working memory research has also seen prominent development in the last few decades. Some of the areas of research have been neural substrates of working memory development in the early years, the development and implementation of neurocomputational models to understand this development, and also use of such models for intervention purposes (Spencer, 2020).

The role of sleep and its effect on memory has been another prominent area of research. Neuroimaging techniques at different levels, such as functional and structural, are enabling scientists to dig deeper into sleep and investigate its role in memory consolidation and learning. In a recent review of sleep studies, Hoedlmoser et al. (2022) presented the trajectory of sleep-dependent memory processes across the life span and information processing and learning during sleep. Further areas of research in this field are the role of sleep in emotional and cognitive functioning, physiological correlates, and theoretical findings to substantiate the contribution of sleep in learning and memory and age-related changes in brain structure and sleep and their impact on memory.

A body of research has also focused on translational research in which researchers have focused on the real-time time impact of their research. Perri et al. (2019) have reviewed associative memory research in some of the common disorders. Translational research is widely needed in different disorders to increase the precision in diagnosis, prognosis, and intervention design.

Moreover, another dimension of investigation in this research category involves discovery of early-stage biomarkers and neuroimaging indicators, developing effective assessment methods, and enhancing validation of therapies for memory dysfunction. In order to have more efficient intervention techniques, a section of research has also been focusing on the use of neuromodulation methods, i.e., the use of non-invasive brain stimulation techniques such as deep brain stimulation, anodal transcranial direct current stimulation (TDCS), and repetitive transcranial magnetic stimulation (rTMS).

Search for neural, subjective, and other biomarkers in the aging population for early detection of memory disorders such as dementia and Alzheimer's disease has been another important area of research. We will study about these in further detail in Chapters 12 and 13 on memory disorders, respectively.

Advancement in technology has also adversely affected memory processing. Studies have been focusing on investigating the long-lasting effects of excessive video-gaming, uses of smartphones, social media, and other relevant devices on encoding, retrieval, and consolidation of memory. On the other hand, researchers have also been focusing on developing technical tools which can make the detection, intervention, and therapeutic part of neurovegetative disorders and memory dysfunction more adequate.

Memory engram is another prominent area of research. Engrams are a set of neurons in different brain regions which accommodate the anatomical and physiological changes in the brain that occur during encoding, consolidation, or retrieval of any information. Studies have been identifying the methods of engram formation and how they function. There is still a lot more to be explored before we can present an exclusive understanding of how engrams play a role in memory storage and retrieval. The fundamental questions which are yet to be explored are what kind of information are stored in engram and how are they encoded (Miry et al., 2021). Also, the functional and structural changes that occur while engrams are made is to be investigated further.

References

Alvarez, P., & Squire, L. R. (1994). Memory consolidation and the medial temporal lobe: A simple network model. *Proceedings of the National Academy of Sciences*, *91*(15), 7041–7045.

Baddeley, A. (2010). Working memory. *Current Biology*, *20*(4), R136–R140.

Binder, J. R., & Desai, R. H. (2011). The neurobiology of semantic memory. *Trends in Cognitive Sciences*, *15*(11), 527–536.

Bruel-Jungerman, E., Davis, S., & Laroche, S. (2007). Brain plasticity mechanisms and memory: A party of four. *The Neuroscientist*, *13*(5), 492–505.

Camina, E., & Güell, F. (2017). The neuroanatomical, neurophysiological and psychological basis of memory: Current models and their origins. *Frontiers in Pharmacology*, *8*, 438.

Christodoulou, C., DeLuca, J., Ricker, J., Madigan, N., Bly, B., Lange, G., Kalnin, A., Liu, W., Steffener, J., & Diamond, B. (2001). Functional magnetic resonance imaging of working memory impairment after traumatic brain injury. *Journal of Neurology, Neurosurgery & Psychiatry*, *71*(2), 161–168.

Eichenbaum, H. (2016). Still searching for the engram. *Learning & Behavior*, *44*, 209–222.

Fish, J., Wilson, B. A., & Manly, T. (2010). The assessment and rehabilitation of prospective memory problems in people with neurological disorders: A review. *Neuropsychological Rehabilitation*, *20*(2), 161–179.

Gazzaley, A., Cooney, J. W., Rissman, J., & D'esposito, M. (2005). Top-down suppression deficit underlies working memory impairment in normal aging. *Nature Neuroscience*, *8*(10), 1298–1300.

Greenwald, A. G., McGhee, D. E., & Schwartz, J. L. (1998). Measuring individual differences in implicit cognition: The implicit association test. *Journal of Personality and Social Psychology*, *74*(6), 1464.

Hebb, D. O. (1949). *The organization of behavior; a neuropsychological theory* (pp. xix, 335). Wiley.

Hoedlmoser, K., Peigneux, P., & Rauchs, G. (2022). Recent advances in memory consolidation and information processing during sleep. *Journal of Sleep Research*, *31*(4), e13607. https://doi.org/10.1111/jsr.13607

Kafkas, A., Mayes, A. R., & Montaldi, D. (2020). Thalamic-medial temporal lobe connectivity underpins familiarity memory. *Cerebral Cortex (New York, N.Y.: 1991)*, *30*(6), 3827–3837. https://doi.org/10.1093/cercor/bhz345

Kalat, J. W. (2015). *Biological psychology*. Cengage Learning. https://books.google.co.in/books?id=EzZBBAAAQBAJ

Kent, P. (2013). The evolution of the Wechsler Memory Scale: A selective review. *Applied Neuropsychology: Adult*, *20*(4), 277–291.

Khalaf, O., Resch, S., Dixsaut, L., Gorden, V., Glauser, L., & Gräff, J. (2018). Reactivation of recall-induced neurons contributes to remote fear memory attenuation. *Science, 360*(6394), 1239–1242.

Kumar, A. A. (2021). Semantic memory: A review of methods, models, and current challenges. *Psychonomic Bulletin & Review, 28*(1), 40–80. https://doi.org/10.3758/s13423-020-01792-x

Kvavilashvili, L., & Ellis, J. (1996). Varieties of intention: Some distinctions and classifications. In M. A. Brandimonte, G. O. Einstein, & M. A. McDaniel (Eds.), *Prospective memory: Theory and applications* (1st ed., pp. 183–207). Psychology Press. https://doi.org/10.4324/9781315806488

Lashley, K. S. (1950). In search of the engram. In Society for Experimental Biology, *Physiological mechanisms in animal behavior* (Society's Symposium IV.) (pp. 454–482). Academic Press.

Mateos-Aparicio, P., & Rodríguez-Moreno, A. (2019). The impact of studying brain plasticity. *Frontiers in Cellular Neuroscience, 13*, 66.

McDaniel, M. A., Glisky, E. L., Guynn, M. J., & Routhieaux, B. C. (1999). Prospective memory: A neuropsychological study. *Neuropsychology, 13*(1), 103.

McGaugh, J. L. (2000). Memory—A century of consolidation. *Science, 287*(5451), 248–251.

Miller, G. A. (1956). The magical number seven, plus or minus two: Some limits on our capacity for processing information. *Psychological Review, 63*(2), 81.

Miry, O., Li, J., & Chen, L. (2021). The quest for the hippocampal memory engram: From theories to experimental evidence. *Frontiers in Behavioral Neuroscience, 14*, 632019. https://doi.org/10.3389/fnbeh.2020.632019

Moradi, A. R., Doost, H. T. N., Taghavi, M. R., Yule, W., & Dalgleish, T. (1999). Everyday memory deficits in children and adolescents with PTSD: Performance on the rivermead behavioural memory test. *The Journal of Child Psychology and Psychiatry and Allied Disciplines, 40*(3), 357–361.

Perri, R., Monaco, M., Fadda, L., Caltagirone, C., & Carlesimo, G. A. (2019). Influence of controlled encoding and retrieval facilitation on memory performance of patients with subcortical ischemic vascular dementia and Alzheimer's disease. *Journal of Neurology, 266*(10), 2447–2456. https://doi.org/10.1007/s00415-019-09411-z

Peterson, L., & Peterson, M. J. (1959). Short-term retention of individual verbal items. *Journal of Experimental Psychology, 58*(3), 193.

Reynolds, C., & Bigler, E. (1994). *Test of memory and learning (TOMAL)*. Pro-Ed.

Roediger, H. L., Weldon, M. S., Stadler, M. L., & Riegler, G. L. (1992). Direct comparison of two implicit memory tests: Word fragment and word stem completion. *Journal of Experimental Psychology: Learning, Memory, and Cognition, 18*(6), 1251.

Ros, L., Romero, D., Ricarte, J. J., Serrano, J. P., Nieto, M., & Latorre, J. M. (2018). Measurement of overgeneral autobiographical memory: Psychometric properties of the autobiographical memory test in young and older populations. *PloS One, 13*(4), e0196073.

Rullier, L., Matharan, F., Barbeau, E. J., Mokri, H., Dartigues, J.-F., Pérès, K., & Amieva, H. (2014). The DMS 48: Norms and diagnostic proprieties for Alzheimer's disease in elderly population from the AMI cohort study. *Gériatrie et Psychologie Neuropsychiatrie Du Vieillissement, 12*(3), 321–330.

Spencer, J. P. (2020). The development of working memory. *Current Directions in Psychological Science, 29*(6), 545–553.

Steffens, D. C., McQuoid, D. R., Welsh-Bohmer, K. A., & Krishnan, K. (2003). Left orbital frontal cortex volume and performance on the benton visual retention test in older depressives and controls. *Neuropsychopharmacology, 28*(12), 2179–2183.

Thompson, R. F. (1986). The neurobiology of learning and memory. *Science, 233*(4767), 941–947.

Tulving, E. (1972). Episodic and semantic memory. *Organization of Memory, 1*(381–403), 1.

Underwood, B. J., & Postman, L. (1960). Extraexperimental sources of interference in forgetting. *Psychological Review, 67*(2), 73.

Williams, J. M. (1991). Memory assessment scales. *Odessa, FL: Psychological Assessment Resources, 199*(1), 1–127.

Wilson, B., Evans, J., Alderman, N., Burgess, P., & Emslie, H. (1997). Behavioural assessment of the dysexecutive syndrome. *Methodology of Frontal and Executive Function, 239,* 250. *Supplementary Materials.*

7 Introduction to Visuospatial Systems

Introduction

Vision is one of the most important senses that we use in our everyday lives. Evolutionarily speaking, the sense of smell was used a lot more than any other sense. However, as we advanced and moved into communities and civilizations, the vision started taking precedence over the others. When we get into the discussion about senses, the first process we need to understand is how the brain perceives its surroundings. The entire process is an exchange of information, and we are only able to perceive something if the object in question is transmitting the information. This information comes in many different forms, and it must be converted into a form appropriate for the brain before it can make sense of it.

An example of this is that the computer stores all its information as a series of 1s and 0s. Anything and everything that goes into a computer or is displayed on a computer are these endless codes of numbers. However, we do not perceive it as such, as the computer translates it in a way that is more suited for our systems. Our brain stores and processes all information as a series of nerve impulses firing together at the same time. René Descartes believed that when we perceive an external stimulus in our environment, it forms a replica of it in our brain. It was Johannes Muller in 1838 who gave the law of specific nerve energies. It was to explain how the same nerve impulse or action potential could be used to transmit different types of information. Different senses activate differently and leave us with various processes.

Visuospatial functioning refers to the ability to process visual or non-verbal information. The dorsal and ventral pathways have been used to integrate the dynamics of the visual-processing systems. Most of the significant overlap of the systems comes through the medial temporal lobe (MTL). This overlap showcases that there is a high interconnection between the two systems. There are areas where there is attention that places a significant role in the modulation of the systems. Attention works in both cases where there are top-down and bottom-up processes. Since, top-down processing is driven by internal experiences, hence top-down (endogenous attentional control) is what is expected by the observer; whereas bottom-up processing is influenced by factors external to

DOI: 10.4324/9781032640839-9

the observer, hence; bottom-up (exogenous attentional control) is the summation of essential physical characteristics in the visual display. This system means that attention is easily directed to the goal and also open to other stimuli (Trés & Brucki, 2014). Vision and visuospatial processing might seem to be the purview of the occipital lobe itself, but other lobes are also involved in the process. The frontal lobe, in particular, is good at eye movements and higher processing in terms of visuospatial working memory. There is no one major center of processing within the brain that is responsible for visual processing. Several systems are formed that are then locally responsible for faces, people, and objects. Visual processing depends on various aspects, but these are more likely to be separated by systems and other primary visual areas. Visual processing depends on various areas, and these are then used to work under many conditions that do not seem to affect the perception itself. The conditions and the circumstances are more liable to changes but the perception of the stimulus remains the same.

The Neurobiology and Physiology of Visual-Spatial Systems

The Eye and Its Connections

The eye functions like a camera, or it would be more prudent to say that it works the other way around. Light, which is the transmission of the stimulus, enters the eye through the pupil, the dark part in the middle of the colored iris. The lens and the cornea then focus the light so it falls on the retina, the portion directly behind the lens. The retina is like a wall filled with visual receptors. These are instrumental in coding the information through the nervous system so the brain can interpret it.

In tracing the path that the information is taken from the stimulus to finally being interpreted as the stimulus, we will encounter all the parts that make up the visual system. There is a select type of neuron that is unique to the retina, known as the bipolar neuron. These then further transmit the information to ganglion cells that take the information directly to the brain. All of these various cells are concentrated near the center of the retina. In order to retain the large surface area afforded by the retina, the axons (wires) of these cells are all concentrated at a point in the retina, leading to the brain. In the absence of any receptor at this spot, this becomes the blind spot in our field of vision. If something in our vision happens to fall in this spot, would the brain perceive it to be non-existent? Not quite. Very often, this space is not big enough to cause any issues. The missing information is usually filled with contextual material.

The eyes and, by extension, the retina is the first stage in primary visual processing. The eye is a special and unique camera that is not yet adequately replicated in the world of technology. There are two types of photoreceptors within the retina which are the rods and the cones. These are involved in the transduction of the electromagnetic wavelengths of light energy and infer the properties of the objects. Cones are instrumental in color perception and are fewer and tend to be situated near the middle of the retina, while the rods are attuned to shades of grey and lead to better vision at night. The retina works contralaterally as well. The right side of the space tends to activate receptors on the left side of the retina, while the left visual field is more likely to activate the right side in this regard. The optic nerves that then carry this information to the brain do so to at least ten different areas in the brain.

Some of these are the pineal gland and the suprachiasmatic nucleus, notable for their role in regulating biological and circadian rhythms. An exciting part of the processing of the information from the retina is that while the right field is processed by the left hemisphere, the image is also reversed in terms of the top and bottom of the image. Visual information reaches the occipital lobes after it undergoes the early stages of processing within the eye (Zillmer et al., 2007). When it comes to the anatomy of the occipital lobe, there are significantly lesser demarcations as compared to other lobes. There are three types of cells in this lobe which are known as the striate, parastriate, and peristriate, based on their striped appearance when they are sectioned. Each level of the lobes shows a precise spatial mapping of the region based on the types of cells. This differentiation would also make it seem that there are possible areas that are mainly responsible for the stimuli, the level, and the space within which it is being processed. The Brodmann area 17, which is also the striate region, would encompass the primary visual cortex. From this area, the information is then moved along to the parastriate region, area 18 that surrounds it. This area is the secondary visual cortex that has rich connections between the hemispheres and through the posterior region of the corpus callosum. The geniculostriate and the tectopulvinar systems are also called the primary and the secondary systems, respectively. The former runs through the lateral geniculate nucleus of the thalamus back to the occipital cortex and takes care of the perception of the forms, patterns, and color. The latter runs through the superior colliculus to the pulvinar and the posterior-lateral regions of the thalamus before spreading to the inferior and middle gyri of the temporal lobes. This system, in particular, looks at the visual location and movement. This division is overly simplified, and the boundaries along anatomy and functions are still not distinct and clear (Beaumont, 2008). The anatomy of the eye and its arrangements shows that complete damage to the optic nerve would mean that there is a kind of blindness (Manly & Mattingley, 2004).

Basic Visual Functions

Following the division of the occipital lobe along with the three types of cells, there are also cases where damage to these areas results in the loss of vision. However, any damage and lesion in the striate cortex would result in a loss of vision. It would only mean an impairment of the visual sensation. Any lesions in the secondary cortex would interfere with visual perception. If there are smaller areas of damage, there are more likely to be gaps in the visual field, known as scotomas that might be surrounded by regions of partial damage, known as amblyopic. Stimulation of the primary visual cortex in patients might result in elementary visual sensations known as photisms.

These are also experienced when struck at the back of the head. An explana-tion of this is that the brain is pushed against the inside of the skull, which leads to mechanical stimulation. A blow from the front could also result in similar photisms. Another aspect of this blindness is a distinct lack of aware-ness regarding the loss of vision. It is more common in cases where there is a progressive deterioration of vision, as in the case of a lesion. One of the primary reasons for this lack of awareness is that of completion. There is a powerful mechanism that seems to construct normal vision based on available visual information. This phenomenon is especially demonstrated in the filling up of the blind spot in the retina. There usually is no awareness of the blind spot as it is usually covered. It would explain why scotomas are also usually ignored. The second factor is that there is just a denial of the disability itself. There is a possibility that the disability was long-standing enough that there was no awareness when the visual deficit started setting in (Beaumont, 2008). A puzzling phenomenon is that patients can respond to visual stimuli that are presented in areas that have been damaged in the striate cortex. The ability of these people to look and point at stimuli without actually seeing is known as blindsight. Blindsight has demonstrated that there is a possibility of training leading to a recovery of the senses based on the increased sensitivity over time (Cowey & Stoerig, 1991).

Higher Visual Processing: Object Recognition and Spatial Localization

This section will deal with the understanding of how higher levels of process-ing occur on visual stimuli. It is where there are more considerable overlaps with the other areas and lobes. There are two streams of visual processing that are differentiated based on the different functions that they provide. The first one is the "what" system, which is also known as the ventral processing system. The second one is the "where" system, also known as the dorsal pro-cessing stream of object localization. Both of these streams of processing are coordinated and integrated through the thalamus. The two streams hypothesis was provided in the face of mounting evidence that there were different neu-ral substrates for the visual processing of action and perception. The ventral stream extends from the striate cortex to the inferotemporal cortex, and the dorsal stream reaches the posterior parietal lobe, according to the Ungerleider and Mishkin model of 1982. This model was then improved later, where the emphasis was not on the distribution between the two streams but on their specialized functions. Patients with visual agnosia following brain damage to the occipitotemporal region often cannot recognize or describe common everyday objects. At the same time, their ability to traverse the world does not seem to be impaired in the slightest. Some patients suffer damage to the optic ataxia, for example, and then their ability to reach visual targets is dam-aged, whereas their recognition ability remains unharmed (Goodale & Milner, 1992). One of the ways there is clarity concerning the different areas of the brain is usually through cases where there is damage to the brain areas. Only when a particular area of the brain is not working does the subsequent behavior or cognitive deficit lead to the identification of that area's exact function.

What and Where of Visual Processing and Its Disorders

The ventral system is specialized for the higher aspects of visual object recognition. It forms the bridge between the visual perception of an object and its recognition. The left hemisphere is more attuned to recognizing symbolic objects like letters and numbers, while the right hemisphere deals with the global identification of objects and faces. Damage to this system can lead to visual agnosia and its two types of apperceptive and associated agnosia.

Agnosias are disorders where a patient is unable to recognize sensory stimuli after suffering brain damage. When differentiating between the two types of agnosias as related above, apperceptive agnosia demonstrates deficits in perceptual processing. In contrast, associative agnosia means that there are no problems or does to a degree where it does not negatively impact the process. There are many subtypes to agnosia where the apperceptive and associative agnosia are basically on a continuum of the deficit level. The types of visual agnosias are visual object agnosia (cannot recognize the meaning of visually presented objects), simultanagnosia (difficulty in understanding the overall meaning of a picture), prosopagnosia (difficulty in recognizing faces), color agnosia, and optic aphasia (difficulty in recognizing visual stimulus, but can be pointed out). The type of agnosia affected is also determined by the localization of the lesion on the brain. An example of this is that apperceptive agnosias are a result of lesions of the sensory association cortex, while associative agnosias are due to lesions of the corticocortical pathways or impairment in areas where semantic representations are stored (Bauer, 2006).

The dorsal system is vital for its role in visually locating and relating. It is the relative distance between other objects and ourselves. There is reciprocal feedback which also helps in planning and coordinating motor movements. Damage to this system leads to right-left discrimination problems, constructional apraxia (CA), and neglect.

Gormley and Brydges (2016) have elaborated on the left-right discrimination difficulty, which is often a result of the dysfunction of various systems like visuospatial processing, memory, language, and the final integration of sensory information. They believe this problem could also be due to the inherent cerebral asymmetry. While this might not seem to be of significant consequence in everyday lives, there are professions like the aviation industry and healthcare where this would have disastrous repercussions. Gainotti and Trojano (2018) have deemed CA to be a typical sign of a parietal lobe lesion. CA is a heterogenous conceptualization that taps into the various systems of the parietal lobe, like visuospatial, perceptual, attentional, planning, and motor mechanisms. All of these must be taken into consideration as these link to the overall pathology of CA. The purview of this disorder is so vast that it encompasses all aspects like assembly, drawing, and building complex models, which are only slightly interconnected. Any damage or lesions on the parietal lobe might manifest in the form of CA. Therefore, CA is loosely defined as any form of difficulty in these tasks that might hamper other functionality as well.

The final and the most widely studied disorder is visual neglect. Certain types of damage alter the overall functionality of the brain. People with neglect lose conscious awareness of a specific aspect of spatial or personal space despite adequate sensory or motor functions. This form of neglect might extend to the motor functions of an entire side where even if the left side of the body is capable of feeling pain, the brain is ignoring it as existent in other regards. There is a lack of conscious awareness of one entire side of the body. This neglect is also demonstrated in the drawings of the patients. There is a focus on only one side of the picture, whereas the other side is not acknowledged at all. The neuropathology of neglect shows that it is most likely to occur where there are lesions in the right inferior parietal lobe or the general posterior regions of the cortex, i.e., the occipital lobe. Other plausible areas are the hemispheres or even the thalamus. Neglect can also manifest in cases where the patient has suffered a stroke or a traumatic brain injury (TBI). There are also cases where the aftermath of electroconvulsive therapy would issue a brief form of neglect. Similar to CA, neglect is also a syndrome that encompasses various aspects of functionality (Zillmer, Spiers, & Culbertson, 2007). A study that sought to understand the anatomy of neglect used a high-resolution MRI protocol to understand the potential mechanisms behind it. In particular, in patients with middle cerebral artery stroke, the primary area involved was the angular gyrus of the inferior parietal lobe. The other area that had not been discussed in previous studies was the parahippocampal region, especially in patients who suffered from a posterior cerebral artery stroke (Mort et al., 2003).

Cloutman (2013) tried to make a case for deciding where the two streams interact and the areas where these are more likely to be differentiated. Both these streams emerge from the early visual areas that are labeled as V1, V2, and V3. These areas are the source that kickstarts the whole process. Entry to the ventral stream is open from the V4 area, while the dorsal stream begins from V5. Cognition is an ongoing process that thrives from the interaction between various modules and activities across the brain for its tasks. There are various types of information that are likely to be needed if the correct usage or information needs to be attended to. These regions are so highly specialized that on their own, there is no way that they would be able to accomplish the complicated task that has been directed.

An example of this would be that while the visual-semantic processing in the ventral stream would name an object like a hammer, the visuomotor processing would give the information with regards to the object's position concerning us. Hence, the integration of both these types of information would be necessary in order to pick up the hammer and use it for its intended purpose. While there is a strong case for the integration of these two streams, there is no idea with regard to where this integration could occur. There are many regions where there are anatomical overlaps between the two pathways, especially in the inferior parietal and inferior temporal regions.

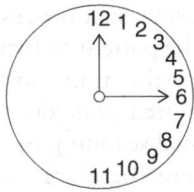

Figure 7.1 Example of a drawing of a clock by a person with contralateral neglect. Retrieved from https://www.neuroscientificallychallenged.com/glossary/contralateral-neglect.

Assessment

Some of the commonly used tests and techniques for the assessment of visuospatial skills are discussed in this section. These are important as corroborating evidence to the imaging evidence and also serve as a confirmation of the clinician's expertise and initial diagnosis. Concerning the assessment of visuospatial functions in people with Parkinson's disease, there are MMSE and specifically for the visuospatial function is the Hooper Visual Organization Test (HVOT) that assesses the ability to rearrange conceptual figures that have been cut into a variety of separate pieces. The target picture must be identified based on the analysis of the set of pieces. The clock drawing test is also useful as it assesses CA, planning, and visuospatial attention as well as neglect. The first part of the test measures the performance of the patient's ability to draw a clock with numbers and set the hands to 1:45. The second part of the test is when the examiner draws a clock and sets the time, and the patient is allowed to copy this clock (Pena et al., 2008) (Figure 7.1).

The object assembly subtest of the Wechsler Adult Intelligence Test – features visuospatial like integration and planning. Individual pieces are provided to the client, and using these, a complete shape or an object must be constructed.

Along with the completion of the task, the performance is also timed, and this becomes another criterion for the severity of the deficit. Another subtest that can be used for this is the block design test. It is a measure of visuoconstructional ability. Tests like the Rey-Osterrieth Complex Figure (ROCF) test and the Bender-Gestalt test have been previously discussed in other chapters (Tyagi & Bhargava, 2011). The Judgment of Line Orientation (JLO) test evaluates the visuospatial skills by asking the patients to match a pair of lines and estimate their incline by comparing them to a reference page that contains 11 lines. The number of correct responses is recorded with a maximum score of 30 (Montse et al., 2001).

Rehabilitation and Treatment

Visuospatial functioning is a vast area, and not all aspects might lend themselves to natural cures and rehabilitation. Some techniques are direct retraining stimulation that remediates basic visual processing. There are visual scanning tasks that can be used to train the patient to focus on the field of neglect. Along with this, other tasks, like size estimation and body awareness, improve visual perception. Depending on the area of neglect, a bright visual stimulus might also be employed to force a visual scanning of the area. Another approach uses a multi-sensory treatment where there are sentences to be copied and auditory cues to scan the neglected hemispaces. Finally, there are repeated cues and practices on the drawing that leads to improvement with each successive round. Most of the assessments can be used as rehabilitation and treatment measures as well (Tyagi & Bhargava, 2011). Cicerone et al. (2005) demonstrated that scanning had been the most effective with visuoperceptual deficits after a right hemispheric stroke. Other forms of training are cued limb activation in the neglected hemispace and voluntary trunk rotation. The recommended guidelines for maximum efficiency are 20 one-hour sessions over four weeks with a colossal intervention apparatus that challenges the boundaries of the neglected hemispace. All of these techniques force the individual to locate and acknowledge the neglected area. Apart from these traditional techniques, transcranial magnetic stimulation or transcranial direct current stimulation has also been increasingly used as a neurorehabilitative tool. Classes of drugs such as dopaminergic and noradrenergic also have potential benefits. Finally, virtual reality produces an environment that is rich and entirely under the control of the clinicians, facilitating better therapeutic outcomes. These are proof of new areas that are untapped as of now. Hence, the research could be directed to increasing the efficacy of the current methods (Fasotti & Van Kessel, 2013). Visuoperceptual disruptions are frequent in stroke patients. Eye-tracking technology that employs cues to signal the patient that they are neglecting a particular field is among the top-rated treatments for neglect. Sensory stimulation has also been proven to show relief. However, these methods are plagued with lots of shortcomings, and none of these is wholly proven as capable treatment modules (Proto et al., 2009). Stroke is one of the major causes of visuospatial deficits. Global-to-local encoding improves retrieval in damaged areas. This encoding strategy involved breaking the figure into individual subunits that were then traced, consequently forcing the individual processing to contribute to the overall figure. The other method employed was the rote repetition technique, which did not lead to similar results, where the entire figure was given at once. The stimulus for the same was the ROCF. There was a significant improvement in visuospatial memory deficits (Chen et al., 2012).

Neurofeedback Applications

Traditionally, neurofeedback is not used for various cognitive deficits and has been somewhat limited in its functions for other aspects. Its efficacy was explored in treating visuospatial neglect to determine the extent of changes. The alpha-rhythm protocol demonstrated a significant correlation between visuospatial deficits and the establishment of alpha-rhythm dynamics. The stability of alpha oscillations is promising in fixing attentional deficits. Alpha amplitude could also be a clinical marker for visual neglect (Ros et al., 2017). This finding was strengthened by using an auditory real-time functional magnetic resonance imaging neurofeedback. It was used to conduct up-regulation of the visual cortex, which resulted in a mild reduction of neglect. This result has been offered as a proof of concept that neurofeedback may be a novel approach to rehabilitation of hemineglect symptoms (Robineau et al., 2019). The treatment seems restricted to unilateral neglect and improving visual attention deficits.

References

Bauer, R. M. (2006). The agnosias. In P.J. Snyder, P.D. Nussbaum, & D.L. Robins (Eds)., *Clinical neuropsychology: A pocket handbook for assessment* (2nd ed., pp. 508–533). American Psychological Association.

Beaumont, J. G. (2008). *Introduction to neuropsychology*. Guilford Press.

Chen, P., Hartman, A. J., Priscilla Galarza, C., & DeLuca, J. (2012). Global processing training to improve visuospatial memory deficits after right-brain stroke. *Archives of Clinical Neuropsychology : The Official Journal of the National Academy of Neuropsychologists, 27*(8), 891–905. https://doi.org/10.1093/arclin/acs089

Cicerone, K. D., Dahlberg, C., Malec, J. F., Langenbahn, D. M., Felicetti, T., Kneipp, S., Ellmo, W., Kalmar, K., Giacino, J. T., Harley, J. P., Laatsch, L., Morse, P. A., & Catanese, J. (2005). Evidence-based cognitive rehabilitation: Updated review of the literature from 1998 through 2002. *Archives of Physical Medicine and Rehabilitation, 86*, 1681–1692.

Cloutman, L. L. (2013). Interaction between dorsal and ventral processing streams: Where, when and how? *Brain and Language, 127*(2), 251–263.

Cowey, A., & Stoerig, P. (1991). The neurobiology of blindsight. *Trends in neurosciences, 14*(4), 140–145.

Fasotti, L., & Van Kessel, M. E. (2013). Novel insights in the rehabilitation of neglect. *Frontiers in human neuroscience, 7*, 780. https://doi.org/10.3389/fnhum.2013.00780

Gainotti, G., & Trojano, L. (2018). Constructional apraxia. In G. Vallar & H. B. Coslett (Eds.), *Handbook of clinical neurology* (Vol. 2.) *The parietal lobe*. Elsevier.

Goodale, M. A., & Milner, A. D. (1992). Separate visual pathways for perception and action. *Trends in Neurosciences, 15*, 20–25.

Gormley, G., & Brydges, R. (2016). Difficulty with right-left discrimination: A clinical problem? *CMAJ: Canadian Medical Association Journal = journal de l'Association medicale canadienne, 188*(2), 98–99. https://doi.org/10.1503/cmaj.150577

Manly, T., & Mattingley, J. B. (2004). Visuospatial and attentional disorders. In L. H. Goldstein & J. E. McNeil (Eds.), *Clinical neuropsychology: A practical guide to assessment and management for clinicians* (Vol. 2, pp. 261–291). Wiley.

Montse, A., Pere, V., Carme, J., Francesc, V., & Eduardo, T. (2001). Visuospatial deficits in parkinsons disease assessed by judgment of line orientation test: Error analyses and practice effects. *Journal of Clinical and Experimental Neuropsychology, 23*(5), 592–598. https://doi.org/10.1076/jcen.23.5.592.1248

Mort, D. J., Malhotra, P., Mannan, S. K., Rorden, C., Pambakian, A., Kennard, C., & Husain, M. (2003). The anatomy of visual neglect. *Brain*, *126*(9), 1986–1997. https://doi.org/10.1093/brain/awg200

Pena, M., Sobreira, E., Souza, C. P., Oliveira, G. N., Tumas, V., & do Vale, F. (2008). Visuospatial cognitive tests for the evaluation of patients with Parkinson's disease. *Dementia & Neuropsychologia*, *2*(3), 201–205. https://doi.org/10.1590/S1980-57642009DN20300007

Proto, D., Pella, R. D., Hill, B. D., & Gouvier, W. (2009). Assessment and rehabilitation of acquired visuospatial and proprioceptive deficits associated with visuospatial neglect. *NeuroRehabilitation*, *24*(2), 145–157. https://doi.org/10.3233/NRE-2009-0463

Robineau, F., Saj, A., Neveu, R., Van De Ville, D., Scharnowski, F., & Vuilleumier, P. (2019). Using real-time fMRI neurofeedback to restore right occipital cortex activity in patients with left visuo-spatial neglect: Proof-of-principle and preliminary results. *Neuropsychological Rehabilitation*, *29*(3), 339–360.

Ros, T., Michela, A., Bellman, A., Vuadens, P., Saj, A., & Vuilleumier, P. (2017). Increased alpha-rhythm dynamic range promotes recovery from visuospatial neglect: A neurofeedback study. *Neural Plasticity*, *2017*, Article 7407241.

Trés, E. S., & Brucki, S. (2014). Visuospatial processing: A review from basic to current concepts. *Dementia & Neuropsychologia*, *8*(2), 175–181. https://doi.org/10.1590/S1980-57642014DN82000014

Tyagi, S., & Bhargava, R. (2011). Visuo-spatial functions: Assessment & rehabilitation. In V. Sharma & S. Malhotra (Eds.), *Clinical neuropsychology*, 8 (pp. 157–171). Harprasad Institute of Behavioral Science.

Zillmer, E., Spiers, M., & Culbertson, W. (2007). *Principles of neuropsychology*. Nelson Education.

8 Neuropsychology of Executive Function

Introduction

Executive functioning is an essential part of what it is to be a human being. It is the very top of the chain where almost all processes are controlled via the executive functioning mechanism, which is the frontal lobe. It would not be an exaggeration to say that this is the main differentiating factor between humans and other animals or living creatures on Earth. "Executive functions are the high-level cognitive processes that facilitate new ways of behaving and optimize one's approach to unfamiliar circumstances" (Gilbert & Burgess, 2008). Going through this definition, it becomes clear that executive functioning processes are the overarching process that dictates and often facilitates other processes. It is a reasonably creative process as well, which ensures that it draws on all the resources available in the brain. More importantly, it has access to these resources as well. The brain usually has clear answers for situations that have been encountered. The more times this situation arises, the more likely these will have an immediate response. Executive functioning is said to be based on the frontal lobe. The frontal lobe is the last area to develop and is hence the seat for most of the functions that are deemed to be high-level processes. While it could be argued that all of these only deal with high-level processing, it deals with the allocation of resources between the two types of processing. EF is an encompassing term for many processes that are carried out by the frontal lobe. Some of these are verbal reasoning, problem-solving, planning, sustained attention, and cognitive flexibility.

These functions can be easily divided into two types. The first one is the type of cognitive function that usually leads to decisions or ideas that are based on logic. They do not induce any emotional reaction. These cognitive processes are planning, sequencing, organizing, and abstract reasoning. All of these processes involve the dorsolateral prefrontal lobe. The other processes that tend to be emotionally charged, like behavior regulation and decision-making, involve the orbitofrontal and the medial frontal cortex. Concerning executive functioning, an area as heterogeneous as the frontal cortex cannot account for all the processes. The brain is highly interconnected, and it is plausible that there are other areas involved as well. The cerebrum merely mediates

DOI: 10.4324/9781032640839-10

all of these areas. Some of the essential functions of these EFs are goal-directed behavior, planning creative tasks, and adapting to uncertain situations that one finds themselves in. It also regulates and inhibits any unnecessary behavior as well (Kedia, 2011).

Models of Executive Functions

Luria Model for Executive Functions

Luria (1966) was the first to conceptualize a model for executive functioning. He observed the behavior of patients with frontal lobe damage as they attempted tasks of problem-solving. He believed that the main parts of executive functioning were "anticipation (setting realistic expectations, understanding consequences), planning (organization), execution (flexibility, maintaining set), and self-monitoring (emotional control, error recognition)." Each of these processes would have a dedicated circuitry to them, with the frontal lobe playing the primary role of assigning tasks and overall regulation of each of these circuits (García-Madruga et al., 2016).

Supervisory Attentional System

Norman and Shallice (1986) conceptualized a model of EFs that focused on attention. It gave two mechanisms that are split across tasks that are either repetitive and routine or the other which are done for the very first time and would, therefore, require a lot more input and attention. The contention scheduling mechanism operates for routine tasks and chooses from competing schemes, and the supervisory attentional mechanism operates for new actions that have not been done often. As a result, there are three modes in which the EFs can operate: automatic, contention scheduling without deliberate direction, and deliberate conscious control. This model is also a valid explanation for how multi-tasking works in our day-to-day lives. Many tasks work seamlessly together. How often have you caught yourself humming a song while writing an assignment or even while driving and also arguing about which song is better for those drives with the other people in your car? Looking at attention as a limited resource means that the primary function is first to direct this to the immediate tasks at hand and then later also reallocate these as and when the tasks are done. Therefore, it would not be impossible if many tasks have to happen at the same time. Another useful analogy would probably be airports where the runway can only accommodate a certain number of aircraft. It is the job of the air traffic controller to ensure that all the aircraft can use the runway and that these are then assigned once a take-off or landing is done. These are called action sequences in the model, which have many coordinated motor responses and also involve the memory systems of the brain. All of these are organized around schemas that work in tandem once they have been set in motion. Whenever there are action sequences that might conflict, the contention scheduling tends to kick in at that point. In these situations, the action sequences either cooperate or compete for the necessary resource. These are all tasks that would not strain the attention stores and can operate without much conscious effort. However, whenever there are tasks that tend to require more than is available, all the resources are sent to that task. In a study by Cieslik et al. (2015), the key regions essential to supervisory attentional control were identified. These were the anterior insula, inferior frontal gyrus, anterior midcingulate cortex, and the pre-supplementary motor area. Of these, the anterior insula and the inferior frontal gyrus were activated for many tasks that differed across paradigms. It implies that these are overarching circuits that are not restricted by domains of the tasks at hand.

Tripartite Model

This model expanded on three systems that interact with each other to monitor the individual's attention and EFs. These were the anterior reticular activating system (ARAS), the front-thalamic gating system, and the diffuse thalamic projection system. Of these three systems, the gating system is responsible for executive attentional control, while the other two maintain attention. The ARAS, in particular, is responsible for the overall level of arousal of the individual, and any damage to this area would result in loss of consciousness. The diffuse thalamic projection system is concerned with maintaining attention over long periods and would change this according to changes in external stimuli. Lastly, the front-thalamic gating system is responsible for high-level cortical functioning like planning, stimuli, and response selection (Chan et al., 2008; Stuss et al., 1986).

Goal Neglect Theory

The frontal lobe is responsible for most goal-directed activity, and most of the human behavior is conducted with a goal in mind. These goals are responsible for the optimal functioning of the human body. If the frontal lobe gets damaged in any way, it usually manifests itself in disorganized behaviors and neglect of any goal-directed activity. Therefore, it was conceptualized as a model in and of itself. The traditional view is that behavior and goal-directed activity has no real connection with intelligence.

Nevertheless, it has been implicated with the g factor as well as frontal lobe functions. This behavior would be based on internal or external instructions. There would also be constraints that restrict the completion of the tasks. Intelligence was not related to g or the frontal lobe. However, the g and frontal lobes are necessary for new tasks and situations. The paradox of frontal lobe dysfunction is that despite having deficits in planning or organization, the intelligence of patients with frontal lobe damage is still intact. The tendency to neglect goals remains high in this population, which is a significant deficit where executive functioning is concerned (Duncan et al., 1996).

Banich's Cascade of Control Model?

Banich (2009) is the latest attempt in this field to integrate a model that would help explain how each component affects actions. It sets up a sequence and a hierarchy for each of these. The areas that are employed in this model are the posterior dorsolateral prefrontal cortex (p-DLPFC), mid-dorsolateral prefrontal cortex (m-DLPFC), posterior-anterior cingulate cortex (p-ACC), and the posterior dorsal anterior cingulate cortex (pd-ACC). This model attempts to integrate all the models that have come before it. It has identified three levels that operate in EFs: neurobiological, psychological, and computational. Linking the previously identified cortical structures across these levels would be the key to the new model. Looking at this task from the context of the Stroop task, the p-DLPFC makes the top-down attentional set for task-relevant goals. In a Stroop task, it would be to say the color and not the word. The p-DLPFC would lead to the activation of regions in the brain that are connected to color identification. The attempt is to overcome the automatic bias to word reading through the manufactured color bias as a task-relevant goal. The m-DLPFC further attenuates the attention to identifying the color of the word. The pd-ACC determines the information parts for the response with the two colors (word and ink) that form the stimulus in the Stroop task. It also serves to rectify any poor selection from the prior areas. The more anterior parts of the pd-ACC evaluate the response that is given. In case the response is not in line with the re-established task-relevant goals, then it also sends a signal to the p-DLPFC to send a better top-down attentional set.

Dysexecutive Syndrome

These are syndromes that evolve as a result of damage to the frontal lobe structures that support cognitive processes. The main issue with dysexecutive functions is that they are challenging to assess and despite proven dependence on intelligence (Duncan et al., 1996), there are often reserved IQs and unimpaired EF on test and lab experiments. Damage to the frontal lobe leads to changes across all the domains of cognition, emotion, and behavior. The very nature of higher-level EFs means that they are difficult to localize to any particular cortical region. Most of the attribution of planning the organization to the frontal lobe came from the dysfunction that was observed in patients with frontal lobe damage. It is also due to its late evolutionary development as well. This portion forms a superstructure, but it is not the actual source of these functions. When it comes to functions like planning, any damage to the dorsolateral prefrontal cortex impairs this process. However, planning in tasks like the Tower of Hanoi, which is usually used as a means for assessment, induces a separate planning deficit. When completing tasks every day, we are more likely to have coordination between multiple regulatory systems with the executive system. Another difficult one is the failure to adjust behavior appropriately to changes that are happening around. The orbitofrontal cortex (OFC) has been implicated in the decision-making process. Any damage to this area means that there is faulty decision-making where the rewards and punishments are not adequately assessed. This area, along with the ventromedial prefrontal cortex (vmPFC), has also been implicated in gambling tasks. It is extensively connected with areas such as the limbic system and the hippocampus as well (Hanna-Pladdy, 2007).

Assessment

a Types

    i Clinical interview

    Every clinical assessment or interview is necessary for assessing deficits and problems in the patients. It serves as an understanding of how the clients themselves view the problems or the deficits that they might be exhibiting. Since EFs are more prominent in every activity, questions regarding their perception of how well they can carry out these activities would be significant. However, patients would often not have insight into their condition, which necessitates the use of corroborating information through other means.

    ii Neuropsychological assessment

    iii Behavioral/functional assessment

       1 **Dysexecutive questionnaire (DEX)** is a 20-item questionnaire describing behavior associated with the dysexecutive syndrome (Behavioural Assessment of the Dysexecutive Syndrome, BADS). Ratings are based on the frequency with which the particular behavior occurs on a Likert scale.

       It is the battery of tests designed to assess the effects of dysexecutive syndrome. As discussed, these are a cluster of impairments that are associated with damage to the frontal lobe. There are difficulties in high-level tasks like planning, initiating, and adapting behavior. The tests in this battery were developed in response to the concerns that previous tests had low ecological validity. Most neuropsychological tests tend to assess functions in isolation which would not work for EFs as these often work in tandem with other functions and areas of the brain. The battery contains six tests and two questionnaires that require participants to engage in EFs.

       The test enjoys significant reliability in terms of inter-rater reliability of 0.88–1. It has good face validity, and there it also enjoys good construct validity. It also assessed participants on problems that they were likely to encounter in their daily lives. The DEX was analyzed factorially to give the three factors of cognitive, behavioral, and emotional (Wilson et al., 1997).

       2 **Rule shift cards** is similar to the Wisconsin Card Sorting Test (WCST), where there is a test of perseveration and mental flexibility.

Stimuli are red or black cards that must be adequately responded to according to the rules that are presented in the beginning. Emphasis is on speed and shifting the rules.

3 **Action programs** are designed to test the ability to implement new solutions to practical problems that must also work according to a set of rules.

4 **Key search** assesses the ability to plan strategies for a problem. The score is based on many criteria that also include whether the examiner thinks that the solution was efficient and useful.

5 **Temporal judgment** assesses abstract thinking based on shared knowledge and everyday events.

6 **Zoo map** is a tool to test the ability to formulate a plan and follow a plan. There are many routes to be taken on the map that does violate the established rules.

7 **Modified six elements** looks at the ability to manage time. There are several tasks like picture naming, dictation, and arithmetic.

b Neuropsychological assessments

i Controlled Oral Word Association Test

ii Animal naming tests test the semantic fluency of the brain in terms of organizing and responding in coordination with the hippocampus.

iii Design fluency test

iv N-back test (working memory) assesses the capacity of the brain to update and monitor the content, which is temporally presented as sequences.

v Tower Test and Maze Tests (planning)

vi Wisconsin Card Sorting Test (set-shifting). This test involves four sets of cards that are required to be sorted according to a constraint that only the examiner knows about. The participant then has to image out the constraint that the examiner has in mind based only on the feedback given by the examiner. This rule also keeps shifting, and the flexibility and the quickness with which the participant can adapt to the rules is also a criterion.

vii Trail Making Test (A and B) (attention). There are a series of colors that are also followed by numbers. The participant has to ensure that they can follow the trail and find the next number in the trail or the series.

viii Stroop Color Word (response inhibition) Test relies on the original word reading bias that is inherent in us. Individuals are more likely to read the word rather than the color that the word is written in. Hence, the original word response has to be inhibited for the color response. These require selective attention to the colors and ignoring the words.

ix Hayling Sentence Completion measures response instigation and response suppression. The test consists of two sets of 15 sentences, each having the last word missing. In set 1, the examiner reads each sentence aloud, and the participant has to complete the sentences, which measures the speed of initiation. "The old house will be ..." In set 2, the patient is instructed to complete the sentence with something that does not fit the sentence. "The commander wanted to stay with the sinking..." This sentence makes the patient inhibit the active response ship in place of the other response that does not make sense for that exact sentence. The time latencies and the number of errors give the scaled scores for assessing executive functioning. Some of its advantages are that the test is quick to administer and easy to score. The real test is verbally administered so people with illiteracy will also be able to participate. However, there is no sizable normative sample, and it has concurrent variable validity (Burgess & Shallice, 1997).

x Simple reaction time

Rehabilitation

Approaches

Physical therapy approach. More often than not, there are more likely to be motor deficits as a result of executive dysfunction. It looks at how insight formation and awareness would go a long way in handling their clients. It also hastens the rehabilitation process tremendously. The suggested task to accomplish this goal must be challenging enough that they are a challenge but not so much that they tend to backfire if the patients are not able to finish them. A similar approach would also help when assisting patients with problem-solving exercises. More often than not, the rehabilitation of EFs has been a case of compensating for the deficits, usually through mnemonics and the like. However, these tend to be more of a burden and a source of cognitive overload rather than aids. Hence, there has been a shift in the remediation of the issues rather than an overreliance on external tools. It is also essential that some level of neuroplasticity would also grant enough access and autonomy to restore executive functionality. This result would be accurate for both motor and cognitive EFs (Studer, 2007).

Recent Advances in Intervention

One of these recent interventions was designed by Levine et al. (2011), focusing on goals and their management in their goal management training (GMT) intervention. It is a model that derives from theories about how goals are processed and sustains attention. It tries to adopt a more mindful approach so that there is more likelihood of monitoring and adjusting goals as and when they are reached. The right frontal-thalamic-parietal sustained attention system is active in most cases. Whenever this system is said to be damaged, there is no goal following, and cues in the environment may tend to distract the patient quickly. The GMT tries to set goals according to set hierarchies. There are lots of instructional videos and materials for the same. All of the sessions were focused on identifying where the patients were likely to lose track of goals and how they could bring attention back. It also promoted mindfulness and being aware at all times whenever possible.

Research Perspective

Executive function (EF) is a multifaceted concept with a variety of higher order cognitive functions that are required in the planning, execution, re-evaluation, and maintenance of goal-directed behavior. Since EFs are a set of core neuro-cognitive functions, their development trajectory plays a very important role in defining the behavior pattern and pathology of an individual. This has made EFs and their development a core area of research in neuropsychology and neuroscience. Several studies have been done to study the development of EF in early childhood and its impairments in the aging population. A relatively limited number of studies are available where EF in adults of middle age has been studied. Critical challenges also involve the unavailability of a set of tasks applicable to all age groups to study EFs and the associated brain structures (Hughes, 2023). Ferguson et al. (2021) studied working memory, inhibitory control, cognitive flexibility, and planning using a battery of tasks to study the age-related differences in EFs in a sample group aged 10–86. This study adds valuable information to the available literature on the developmental trajectory of different components of EFs. The study has indicated that different cognitive functions follow different developmental trajectories, and their study at different life stages is essential to better understand cognitive aging. Different cognitive functions that come under EFs are needed to be studied across gender and in cross-cultural settings for different age groups to have a comprehensive understanding of changes in executive functioning. These can help in the development of effective interventions for executive dysfunctions.

Hughes (2023) in his commentary on where we stand in our understanding of EFs and whether it needs to be changed, has reflected upon the effect of variation of culture, community, individual difference, the effect of practice AIT intervention, and the age-related developmental changes. Based on the studies he has referred to, he presents that factors like the differences in values and knowledge of the culture, the context in which information is processed in that culture has an impact on which and how a particular cognitive component of EF will be used. This results in varying development of EFs in different cultures and different countries. Furthermore, the performance of children and adults on tasks to assess various EFs will vary in different cultural setups based on knowledge, belief, context, and the use of cognitive function. Therefore, based on recent development in the field, there is a need to re-evaluate the task used to study EFs and make them more efficient in a way that can measure not only the individual differences but also the differences in the context.

Another critical area to reflect in the EF research is the use of advanced non-invasive neuroimaging methods to study the brain area activated in executive functioning. As pointed out by Salehinejad et al. (2021), one of the significant limitations of using neuroimaging methods is the lack of a causal relationship between brain area activation and behavior. They have highlighted the benefits of using non-invasive neuromodulation methods for investigating the physiological changes in the brain underpinning cognitive activities. The two non-invasive methods are transcranial magnetic stimulation (TMS) and transcranial electric stimulation (tES), which change the polarity and cortical activity within the brain area over which applied, which results in changes in brain function.

The literature also distinguishes between different EFs based on the level of involvement of emotions in them. There are two categories: hot (affect and motivation-driven) and cold (purely cognitive) EFs. For example, hot EFs are emotion regulation, reward selection, decision-making and cold EFs are working memory, inhibitory control, reasoning, etc. Based on the task required in the execution of a particular execution function, brain areas are activated. Studies have shown that the prefrontal cortex and associated cortical regions are involved in the processing of EFs. The neural substrate of EF includes the orbital and medial prefrontal cortex and associated regions for the processing of hot EFs, and the lateral prefrontal cortex, including the dorsolateral prefrontal cortex and ventrolateral prefrontal cortex are the regions involved in cold EFs. A large number of neuroimaging studies have shown that for some of the hot and cold EFs, brain region activation overlaps for some of the tasks. Building upon the findings of prior studies, Salehinejad et al. (2021) have presented a new network-based approach to hot and cold categories in EFs as it offers a more efficient way of identifying the pathophysiology underlying neuropsychiatric disorders, which often involve both emotional and cognitive dysfunction. This has the potential for the development of effective therapeutic interventions if aligned well with the cognitive profiles of these disorders. A different study identified the lack of a larger sample size, a comprehensive battery with a set of tasks for different age groups to study hot and cold EFs to provide more generalized and valid results (Poon, 2018).

As it is seen, there are a variety of neuroimaging studies done in the last couple of decades. Still, there is enormous scope for using advanced neuroimaging methods such as functional near-infrared spectroscopy (fNIRS), myelin-specific MRI technique for studying brain networks in infants and young children, with larger sample pool, and longitudinal studies for better identification of the development of EFs can answer so many questions (Fiske & Holmboe, 2019).

A distinct sphere of research to consider is EF across the life span. An extensive systematic review of literature on the age group of 6–12-year-old children identified that three of the most studied components of EF are working memory, cognitive flexibility, and inhibitory control (Fernández García et al., 2021). They also identified the lack of studies on delayed gratification and decision-making. There is a need for longitudinal studies on understanding the developmental trajectories of various hot and cold EFs using newly developing non-invasive brain stimulation methods. The findings from these studies can help in better understanding of cognitive and socio-affective aspects of neurodevelopmental disorders.

References

Banich, M. T. (2009). Executive function: The search for an integrated account. *Current Directions in Psychological Science, 18*(2), 89–94. https://doi.org/10.1111/j.1467-8721.2009.01615.x

Burgess, P. W., & Shallice, T. (1997). *The hayling and Brixton tests.* Thames Valley Test Company.

Chan, R. C., Shum, D., Toulopoulou, T., & Chen, E. Y. (2008). Assessment of executive functions: Review of instruments and identification of critical issues. *Archives of Clinical Neuropsychology, 23*(2), 201–216.

Cieslik, E. C., Mueller, V. I., Eickhoff, C. R., Langner, R., & Eickhoff, S. B. (2015). Three key regions for supervisory attentional control: Evidence from neuroimaging meta-analyses. *Neuroscience & Biobehavioral Reviews, 48*, 22–34.

Duncan, J., Emslie, H., Williams, P., Johnson, R., & Freer, C. (1996). Intelligence and the frontal lobe: The organization of goal-directed behavior. *Cognitive Psychology, 30*(3), 257–303. https://doi.org/10.1006/cogp.1996.0008

Ferguson, H. J., Brunsdon, V. E. A., & Bradford, E. E. F. (2021). The developmental trajectories of executive function from adolescence to old age. *Scientific Reports, 11*(1), 1382. https://doi.org/10.1038/s41598-020-80866-1

Fernández García, L., Merchán, A., Phillips-Silver, J., & Daza González, M. T. (2021). Neuropsychological development of cool and hot executive functions between 6 and 12 years of age: A systematic review. *Frontiers in Psychology, 12*. https://www.frontiersin.org/articles/10.3389/fpsyg.2021.687337

Fiske, A., & Holmboe, K. (2019). Neural substrates of early executive function development. *Developmental Review, 52*, 42–62. https://doi.org/10.1016/j.dr.2019.100866

García-Madruga, J. A., Gómez-Veiga, I., & Vila, J. Ó. (2016). Executive functions and the improvement of thinking abilities: The intervention in reading comprehension. *Frontiers in Psychology, 7*, 58.

Gilbert, S. J., & Burgess, P. W. (2008). Executive function. *Current Biology, 18*(3), R110–R114.

Hanna-Pladdy, B. (2007). Dysexecutive syndromes in neurologic disease. *Journal of Neurologic Physical Therapy, 31*(3), 119–127.

Hughes, C. (2023). Executive functions: Going places at pace. *Journal of Cognition and Development, 24*(2), 296–306. https://doi.org/10.1080/15248372.2023.2187636

Kedia, S. (2011). Executive functions: Assessments and rehabilitation. In V. Sharma & S. Malhotra (Eds.), *Clinical neuropsychology* (1st ed., pp. 157–171). Harprasad Institute of Behavioral Sciences.

Levine, B., Schweizer, T. A., O'Connor, C., Turner, G., Gillingham, S., Stuss, D. T., Manly, T., & Robertson, I. H. (2011). Rehabilitation of executive functioning in patients with frontal lobe brain damage with goal management training. *Frontiers Human Neuroscience, 5*, 9.

Luria, A. R. (1966). *Higher cortical functions in man.* Basic Books.

Norman, D. A., & Shallice, T. (1986). Attention to action. In R. J. Davidson, G. E. Schwartz, & D. Shapiro (Eds.), *Consciousness and self-regulation: Advances in research and theory volume 4* (pp. 1–18). Springer US. https://doi.org/10.1007/978-1-4757-0629-1_1

Poon, K. (2018). Hot and cool executive functions in adolescence: Development and contributions to important developmental outcomes. *Frontiers in Psychology, 8.* https://www.frontiersin.org/articles/10.3389/fpsyg.2017.02311

Salehinejad, M. A., Ghanavati, E., Rashid, M. H. A., & Nitsche, M. A. (2021). Hot and cold executive functions in the brain: A prefrontal-cingular network. *Brain and Neuroscience Advances, 5,* 23982128211007770. https://doi.org/10.1177/23982128211007769

Studer, M. (2007). Rehabilitation of executive function: To err is human, to be aware— Divine. *Journal of Neurologic Physical Therapy, 31*(3), 128–134.

Stuss, D., Benson, D., Clermont, R., Della Malva, C., Kaplan, E., & Weir, W. (1986). Language functioning after bilateral prefrontal leukotomy. *Brain and Language, 28*(1), 66–70.

Wilson, B., Evans, J., Alderman, N., Burgess, P., & Emslie, H. (1997). Behavioural assessment of the dysexecutive syndrome. *Methodology of Frontal and Executive Function, 239,* 250. *Supplementary Materials.*

9 The Neuropsychology of Social Cognition

Introduction

Social cognition emerged as an individual field of study in the late 1980s when cognitive psychologists started exploring the brain mechanisms underpinning social behavior. Until the 1970s, social psychologists explored the interplay between social factors and human behavior and their impact on each other. The social psychologist studied learning, memory formation, theory of self, attribution, and concepts like prejudice, stereotype, and attribution. In contrast, cognitive psychologists worked on understanding possible brain mechanisms underlying human behavior. The quest to understand what cognitive processes within the brain influence social behavior, how such information is being stored and processed in the brain (Sherman et al., 1989), and how the brain makes sense of ourself and our surroundings led to the emergence of social cognition as a field. Social cognition takes contributions from two of the stalwart schools of psychology – social and cognitive psychology and combines their applications to result in a more holistic and nuanced understanding.

While due to the vast scope of this field, there exists a lack of consensus amongst researchers concerning a best all-encompassing definition, social cognition in its simplest form can be understood as "how people think about other people" and how this influences their ideas of self as well as their interactions with others (Wegner & Vallacher, 1977). Social cognition is a multifaceted process involving several mental processes and structures. Attention, or the capacity to focus on and selectively interpret social information, is crucial to social cognition. A conceptual and empirical approach to understanding social psychological topics by investigating the cognitive underpinning of whatever social phenomena are being studied. People are more likely to pay attention to stimuli that are relevant to their aims or interests, which might affect how they perceive and interpret social information (Kunda, 1999). According to (Fiske & Taylor, 2013), social cognition is the study of how people make sense of others and themselves. It focuses on how ordinary people think and feel about people, including themselves.

DOI: 10.4324/9781032640839-11

Historical Roots of Social Cognition

The premise that "the individual is social" has always held a central role within psychology, and even when different theorists from various psychological orientations came up with their own theories of understanding the behaviors of an individual, there remained a constant underlying theme of social and environmental factors within it. Thinking along this perspective, certain schools of thought that serve as antecedents for shaping social cognition as a field of empirical inquiry are discussed below.

Behaviorism

Behaviorism dominated much of 20th-century psychology and built its entire conceptualization of psychology upon observable and empirical methods. Behaviorists hold that it is the stimulus-response (S-R) associations that can help explain all observable behavior, and if these associations are manipulated, they'll result in behavioral changes. The aim of this school of thought, thus, became to identify these fundamental laws which lead to S-R relations and think of ways in which these can be altered to lead to behavioral modifications. Some of the stalwarts who contributed to behaviorism via their research were John Watson, B. F. Skinner, Hermann Ebbinghaus, and so on. Despite its success, however, behaviorism, over the years, was subjected to various criticisms, one of the most prominent ones being its complete ignorance of the mental processes which might play a role between the environmental stimuli and the exhibited response. Regardless, the behaviorist held an individual's social environment at high precedence and claimed that it is this external environment of any individual that determines which aspects of behaviors are learned, maintained, and thus exhibited versus which behaviors are likely to go extinct (operant conditioning).

Cognitive Psychology

From the criticisms of behaviorism in the 1960s and 1970s, a new field of inquiry called cognitive psychology started to gain precedence. This new subfield started to empirically study the underlying mental processes, which are not directly observable but significantly influence the behavioral responses of an individual. This gave rise to what came to be known as S-O-R psychology, where O stands for the organism that comes with its own patterns of mental sets, schemas, and representations, each of which mediates his/her interaction with the environment. Thus, research traditions within this field started to explore hidden processes such as attention, perception, language, memory, and decision-making, and how these processes develop, integrate, and if they can be manipulated to affect behaviors. The most significant advantage of this field of cognitive psychology was that it maintained its roots within the philosophy of scientific empiricism, and thus all their research employed methods that emphasized reliability, control, and generalizability while also opening up avenues to study mental processes which cannot be directly observed.

Social Psychology

While questions pertaining to social psychology have always been inherent within the discipline of psychology since its genesis, one of the major historical catalysts to nudge it into the spotlight as a separate field of study was the Second World War. During this time, major research emphasis was placed on enhancing leadership, ways to successfully persuade the masses, enhancement of group performance within military contexts, and the like. These developments within the army domain paved the way for social psychology researchers to ask even broader questions applicable more to everyday life contexts. Some of the landmark social psychology experiments which were conducted during this time were Milgram's study on obedience (Milgram, 1963), Zimbardo's prison experiment (Zimbardo et al., 1971), Asch conformity experiment (Asch, 1951), and so on. It has to be noted that developments within social psychology were going parallelly with the stream of research which explored underlying cognitive phenomena affecting these socially observed behaviors. For instance, the attribution theory developed by Kelley (1967) explored how individuals make internal and external attributions and how these are contingent upon one's social contexts and circumstances as well as self-representations and schemas. Thus, social and its implicit cognitions were always intertwined within these streams of research and did not exist as exclusive areas of exploration. However, as the broader aim of this field was still to explore "social psychological outcomes," the study of various cognitive processes influencing these outcomes, while not wholly ignored, mainly remained at the margins. It will be precisely understanding and explaining these processes, which will become a significant area of focus for social cognition theories and research.

Emergence of Social Cognition

The growth of these research traditions and methodologies led to a more nuanced and integrated understanding of social phenomena. Research in the 1970s started demonstrating a more holistic approach within its conceptualizations, where social behaviors were now being perceived from the lens of their underlying cognitive functions. For instance, researchers have now started looking into the mechanisms which lead to the formation of stereotypes rather than just trying to understand their content (Hamilton & Gifford, 1976; Rothbart et al., 1979). Similarly, models of impression formation started demonstrating how impressions are organized and recollected rather than just focusing on the observable aspects which emerge as a consequence of having formed an impression (Hastie & Kumar, 1979). One of the implicit assumptions within social psychology always held that people generally know what they're doing and why they are doing it. With the rise in social cognition, however, researchers started to actively explore the underlying processes that influence this non-conscious, automatic information processing. For example, Nisbett & Wilson (1977) experimentally established how individuals tend to make implicit associations without conscious awareness, which significantly influences their choices and evaluations. The implications these findings held were far and wide. They helped psychologists explain and understand a plethora of social behaviors, such as discrimination, and reasons for the development of ingroup-outgroup biases. These studies opened up gates for the exploration of questions that lay somewhere at the intersection of social and cognitive psychology and thus serve as an important link explaining the overlapping phenomenon between the two.

Over the years, with advancements in understanding, researchers now view social cognition as a general conceptual framework and more of a methodological approach that has been applied to understand the cognitive underpinnings of group processes, aggression, interpersonal relationships, attitudes, and prejudices, and so forth, rather than a subfield of social or cognitive psychology. It recognizes the complex interplay between an individual's expectancies, underlying cognitive processes, and their social-environmental interactions, all of which together aids in the reliable prediction of behavior.

Methods in Social Cognition

Social cognition utilizes the methods used in cognitive psychology and neuropsychology to study its subject matter. Research methods such as behavioral experimental studies, surveys, observational studies, neuroimaging such as functional magnetic resonance imaging (fMRI), electroencephalogram (EEG), and electromyogram (EMG) are used in social cognition to understand the neural mechanism underlying behavior.

Behavioral Experiments

Behavioral experiments are a critical methodology in social cognition research as they provide researchers with a controlled setting to investigate the processes and mechanisms underlying social behavior. These typically involve manipulating social stimuli and measuring participants' responses and can be used to investigate how social factors, such as the presence of others or group norms, influence people. For example, researchers might use a social influence paradigm to investigate how the presence of others affects decision-making. Behavioral experiments can also be used to investigate how people categorize others into social groups and the consequences of these categorizations, how people perceive others and make judgments about them, and social cognitive deficits in clinical populations, such as individuals with schizophrenia or autism. For example, social categorization paradigms are used to investigate how group membership affects interpersonal interactions, perception paradigms to investigate how people form impressions of others based on facial expressions, or using social cognition paradigms to investigate how individuals with schizophrenia perceive emotions in others.

Neuroimaging

Neuroimaging techniques, such as fMRI, have become increasingly important in the study of social cognition. These techniques allow researchers to investigate the neural mechanisms underlying social cognition and behavior. Neuroimaging studies have identified brain regions such as the prefrontal cortex, amygdala, and insula that are involved in perceiving facial expression, empathy, altruism, experiencing emotions, etc. fMRI studies have shown amygdala activation when people perceive emotional facial expressions (Mattavelli et al., 2014), activation of anterior cingulate cortex (ACC) when people experience empathy for others (Lockwood, 2016). Neuroimaging studies have also been used to investigate social cognitive deficits in clinical populations, such as individuals with autism or schizophrenia. For example, fMRI studies have shown that individuals with autism have reduced activation in brain regions involved in social cognition when processing social stimuli (Dichter, 2012). Neuroimaging studies have also been used to investigate how social context affects brain activity. Several fMRI studies have shown that the same stimulus can activate different brain regions depending on whether it is perceived as being part of a social or non-social context.

Surveys

Surveys are also an important tool in the field of social cognition as they allow researchers to gather data on individuals' attitudes, beliefs, and behaviors. Surveys can be used to examine attitudes and beliefs about various social phenomena. For example, surveys have been used to examine attitudes toward immigration, prejudice, and intergroup relations. Surveys can be used to measure social cognition, such as social perception, emotion recognition, and theory of mind. For example, surveys have been used to measure social cognition in clinical populations, such as individuals with autism. Surveys can also be used to examine social identity, such as identification with one's gender, race, or ethnicity. For example, surveys have been used to examine how social identity influences attitudes and behaviors. Surveys can be used to explore intergroup relations and perceptions of different social groups. For example, surveys have been used to explore perceptions of different racial and ethnic groups and their impact on intergroup relations.

Qualitative Studies

Qualitative studies are an important tool in the field of social cognition as they allow researchers to gain a deeper understanding of the experiences, perspectives, and attitudes of individuals. Qualitative studies can be used to examine the experiences of marginalized groups, such as individuals from minority racial or ethnic groups, LGBTQ+ individuals, or individuals with disabilities. These studies can provide insights into the unique challenges and barriers faced by these groups and help inform interventions to address them. Qualitative studies can also be used to explore social perceptions and attitudes toward various social phenomena, such as discrimination, prejudice, or social inequality. These studies can provide a rich understanding of the complexities and nuances of these attitudes.

Social Cognition and Neuropsychology

There isn't one specific area of the brain that we can point to and say, "This is where the self and all processing related to the self is located," despite the fact that each of us has knowledge about ourselves that includes our unique characteristics, beliefs, desires, past, location in space, and the awareness that our body is our own. The "social brain," which also seems to be made up of related systems, some of which may be specifically dedicated to social interactions, is activated when we enter the social environment and engage with others. The cognitive process of connecting information, frequently from the outside world to oneself, is known as self-referential processing. Two examples are the ability to ascribe personality features to oneself or recognize remembered events as being one's memories of the past. The posterior cingulate cortex (PCC), medial and lateral parietal cortex, the dorsolateral prefrontal cortex, and ventromedial prefrontal cortex, and these areas have all been linked to self-referential processing. The orbitofrontal cortex (OFC), ACC, and insula, as well as the autonomic nervous system (ANS), hypothalamic-pituitary-adrenal (HPA) axis, and endocrine systems that control bodily states, emotion, and reactivity, all mediate subjective feelings, which also contribute to our sense of self. The temporal lobe is important because memory is an aspect of self-referential processing. The amygdala, along with its connections to the medial prefrontal cortex (PFC), the OFC, and the superior temporal sulcus (STS), as well as the ACC, the fusiform face area (FFA), areas connected to mirror neurons, the insula, the temporal poles, the temporoparietal junction (TPJ), and the medial parietal cortex, are all activated when we attempt to understand other people (Gazzaniga & LeDoux, 2013).

Social Interactions and Development

The prefrontal cortex, which is essential for impulse control, cognitive control, and decision-making, continues to grow throughout childhood and adolescence. This stage of maturation is accompanied by concurrent developmental changes in social behavior, such as an increase in peer-peer relationships marked by a surplus of social play behavior (Blakemore & Choudhury, 2006; Crone & Dahl, 2012; Spear, 2000). The effects of unpleasant experiences and the benefits of positive ones are more likely to be felt in developing brain areas than in fully formed ones. Negative social experiences like neglect or abuse throughout infancy and adolescence raise the likelihood of developing mental diseases, including depression, anxiety, schizophrenia, or substance misuse later in life (Burke et al., 2017). Researchers frequently utilize rats, who are highly sociable animals with neural and behavioral development similar to humans, to explore the effects of negative social experiences on neurodevelopment. Researchers have shown that socially isolated rats during a period similar to infancy and adolescence in humans severely impede the development of their social skills. Social isolation in rats involves denying them social interaction but allowing them visual, auditory, and olfactory contact with other rats. Reduced dopamine and increased serotonin signaling in the PFC, altered PFC function, disrupted synaptic plasticities, and social and behavioral deficits, such as increased aggression, anxiety, and fear (Lukkes et al., 2009), which persist into adulthood, are some of the findings. Even after being resocialized, adult rats still exhibit decreased social functioning.

Phineas Gage and Social Cognition

The damage suffered by railway worker Phineas Gauge led to the initial discovery that the frontal lobes might contribute very specifically to behavior in the social realm. Gage's frontal lobes, including the ventromedial prefrontal cortex, were severely damaged on either side by an unintentional explosion that sent a metal rod through his brain. Before his injury, Gauge was a responsible, obedient, courteous, and socially savvy guy, but afterward, he behaved in a callous, profane, and socially incorrect manner. Until it could be explained in light of patients who had experienced a similar change in personality in more recent times: similar to Gauge, other patients with bilateral ventromedial frontal lobe injury have a drastically reduced capacity for social interaction despite having completely normal profiles on common neuropsychological tests including IQ, language, perception, and memory. According to recent theoretical interpretations, the ventromedial frontal cortices are crucial in linking emotional experience with decision-making in complicated contexts, maybe notably in social situations (Adolphs, 1999).

Amygdala and Social Cognition

Prejudice

Presentation of faces deemed untrustworthy also triggers the amygdala (Winston et al., 2002). Given that the experiment participants were unfamiliar with the faces, this is an example of bias. Numerous imaging paradigms have been used to study racial prejudice, and each time, amygdala activation has been linked to the unconscious fear that is brought on by looking at the face of a person of a different race. For instance, activity in the amygdala was seen when white Americans were presented with the faces of unknown black Americans (Phelps et al., 2000).

Prejudice and Conditioning

Since amygdala is involved in conditioning fear, it is involved in this process. Numerous studies on animals have demonstrated that the amygdala is a component of a system that develops value associations with stimuli, regardless of whether those stimuli are social (LeDoux, 2000). Positive and negative numbers are both acceptable in this system. For instance, the amygdala not only reacts to things that cause fear because they are connected to punishment (negative value) but it also reacts to things connected to food and sex (positive value).

Ventromedial Prefrontal Cortex and Social Cognition: Somatic Marker Hypothesis

Antonio Damasio and other scholars put up the somatic marker theory, which contends that emotional factors influence decision-making and other behavior in general. "Somatic markers" are bodily sensations that correspond to specific emotions, for as the connection between a quick heartbeat and fear or nausea and disgust. The idea is that somatic markers have a significant impact on future choices. The ventromedial prefrontal cortex and the amygdala are hypothesized to process somatic indicators inside the brain. Using the Iowa gambling task as a test subject, the hypothesis has been investigated.

Mirror Neurons and Theory of Mind

Mirror neurons are a special kind of brain cell that reacts the same way when we carry out an activity and when we see another person carry out the same action. All of the initial mirror-neuron research focused on both humans and monkeys performing and observing actions. Motor neurons alone, however, are unable to address some of the most intriguing concerns that mirror neurons pose. Instead, scientists want to know how we interpret other people's emotions and experiences, not only their movements. Keysers et al. (2004) examined "tactile empathy," or how we react to seeing others being touched. He discovered that when the 14 participants – this time, both men and women – were softly touched on the leg using a device like a feather duster, as well as when they looked at images of someone else being touched in the exact location, the somatosensory cortex was activated.

Impaired Social Cognition in Adolescents

Psychopathology may lead to aberrant social cognition. Conditions such as autism spectrum disorder (ASD), attention-deficit/hyperactivity disorder (ADHD), and depression are social cognitive impairments. Prominent features of ASD include social difficulties and deviated social thinking. ASD is also accompanied by lower reciprocity and mentalizing suggesting impairment in social cognition. Adolescents being a critical period of social development, may provide important insights regarding neural mechanisms underlying social cognition. The default mode network comprises the social brain, and its disruption has been implicated in ASD. In adolescents with ASD, a global pattern of underconnectivity was observed in default mode network (DMN) and between DMN and other brain areas. Moreover, individuals with ASD do not go through age-related maturation of DMN. This implies that the neural correlates of social cognition undergo deviant neurodevelopment during adolescence of individuals with ASD (Nair et al., 2020). Adolescents with depression also show deficits in social cognition, including distorted mental representation, biased information processing, memory and attention, poor decision-making, and poor social coping (Kyte & Goodyer, 2008). Dysregulation of the SPIN model and hence impaired processing of social information may lead to the onset of mood and anxiety disorders during adolescence.

References

Adolphs, R. (1999). Social cognition and the human brain. *Trends in Cognitive Sciences, 3*(12), 469–479.

Asch, S. (1951). *Asch conformity experiment.* Author.

Blakemore, S., & Choudhury, S. (2006). Development of the adolescent brain: Implications for executive function and social cognition. *Journal of Child Psychology and Psychiatry, 47*(3–4), 296–312.

Burke, A. R., McCormick, C. M., Pellis, S. M., & Lukkes, J. L. (2017). Impact of adolescent social experiences on behavior and neural circuits implicated in mental illnesses. *Neuroscience & Biobehavioral Reviews, 76,* 280–300.

Crone, E. A., & Dahl, R. E. (2012). Understanding adolescence as a period of social–affective engagement and goal flexibility. *Nature Reviews Neuroscience, 13*(9), 636–650.

Dichter, G. S. (2012). Functional magnetic resonance imaging of autism spectrum disorders. *Dialogues in Clinical Neuroscience, 14*(3), 319–351.

Fiske, S. T., & Taylor, S. E. (2013). *Social cognition: From brains to culture.* SAGE Publications. https://books.google.co.in/books?id=uVJdBAAAQBAJ

Gazzaniga, M. S., & LeDoux, J. E. (2013). *The integrated mind.* Springer Science & Business Media.

Hamilton, D. L., & Gifford, R. K. (1976). Illusory correlation in interpersonal perception: A cognitive basis of stereotypic judgments. *Journal of Experimental Social Psychology, 12*(4), 392–407.

Hastie, R., & Kumar, P. A. (1979). Person memory: Personality traits as organizing principles in memory for behaviors. *Journal of Personality and Social Psychology, 37*(1), 25.

Kelley, H. H. (1967). *Attribution theory in social psychology.* Nebraska Symposium on Motivation.

Keysers, C., Wicker, B., Gazzola, V., Anton, J.-L., Fogassi, L., & Gallese, V. (2004). A touching sight: SII/PV activation during the observation and experience of touch. *Neuron, 42*(2), 335–346.

Kunda, Z. (1999). *Social cognition: Making sense of people*. A Bradford Book. https://books.google.co.in/books?id=BjnOMFtYiwUC

Kyte, Z., & Goodyer, I. (2008). Social cognition in depressed children and adolescents. In P. Fonagy & I. Goodyer (Eds.), *Social cognition and developmental psychopathology* (pp. 201–238). Oxford University Press. https://doi.org/10.1093/med/9780198569183.003.0008

LeDoux, J. E. (2000). Emotion circuits in the brain. *Annual Review of Neuroscience*, *23*(1), 155–184.

Lockwood, P. L. (2016). The anatomy of empathy: Vicarious experience and disorders of social cognition. *Behavioural Brain Research*, *311*, 255–266. https://doi.org/10.1016/j.bbr.2016.05.048

Lukkes, J. L., Mokin, M. V., Scholl, J. L., & Forster, G. L. (2009). Adult rats exposed to early-life social isolation exhibit increased anxiety and conditioned fear behavior, and altered hormonal stress responses. *Hormones and Behavior*, *55*(1), 248–256.

Mattavelli, G., Sormaz, M., Flack, T., Asghar, A. U. R., Fan, S., Frey, J., Manssuer, L., Usten, D., Young, A. W., & Andrews, T. J. (2014). Neural responses to facial expressions support the role of the amygdala in processing threat. *Social Cognitive and Affective Neuroscience*, *9*(11), 1684–1689. https://doi.org/10.1093/scan/nst162

Milgram, S. (1963). Behavioral study of obedience. *The Journal of Abnormal and Social Psychology*, *67*(4), 371.

Nair, A., Jolliffe, M., Lograsso, Y. S. S., & Bearden, C. E. (2020). A review of default mode network connectivity and its association with social cognition in adolescents with autism spectrum disorder and early-onset psychosis. *Frontiers in Psychiatry*, *11*. https://www.frontiersin.org/articles/10.3389/fpsyt.2020.00614

Nisbett, R. E., & Wilson, T. D. (1977). The halo effect: Evidence for unconscious alteration of judgments. *Journal of Personality and Social Psychology*, *35*(4), 250.

Phelps, E. A., O'Connor, K. J., Cunningham, W. A., Funayama, E. S., Gatenby, J. C., Gore, J. C., & Banaji, M. R. (2000). Performance on indirect measures of race evaluation predicts amygdala activation. *Journal of Cognitive Neuroscience*, *12*(5), 729–738. https://doi.org/10.1162/089892900562552

Rothbart, M., Evans, M., & Fulero, S. (1979). Recall for confirming events: Memory processes and the maintenance of social stereotypes. *Journal of Experimental Social Psychology*, *15*(4), 343–355.

Sherman, S. J., Judd, C. M., & Park, B. (1989). Social cognition. *Annual Review of Psychology*, *40*(1), 281–326. https://doi.org/10.1146/annurev.ps.40.020189.001433

Spear, L. P. (2000). The adolescent brain and age-related behavioral manifestations. *Neuroscience & Biobehavioral Reviews*, *24*(4), 417–463.

Wegner, D. M., & Vallacher, R. R. (1977). *Implicit psychology: An introduction to social cognition*. Oxford University Press.

Winston, J. S., Strange, B. A., O'Doherty, J., & Dolan, R. J. (2002). Automatic and intentional brain responses during evaluation of trustworthiness of faces. *Nature Neuroscience*, *5*(3), 277–283. https://doi.org/10.1038/nn816

Zimbardo, P. G., Haney, C., Banks, W. C., & Jaffe, D. (1971). *The Stanford prison experiment*. Zimbardo, Incorporated.

Part 3

10 Traumatic Head Injury and Rehabilitation

Introduction/Overview

The brain is a well-protected organ within the human body. When it comes to the different parts of the body, there are more likely injuries and damages as we grow and evolve. Nearly all of these injuries would go on to heal all on their own with little long-lasting effects. However, when it comes to the brain, traumatic brain injury, or TBI for short, is one of the most distressing things to a brain. The impact of a TBI can range from a simple tap to the back of the head to a gunshot wound to the head. The damage is diffused to various parts of the brain and leads to many localized deficits. This chapter will go on to describe what constitutes a TBI and its epidemiology nationally and globally. We shall also discuss the various types of traumas that get inflicted on the brain as a result. The assessment of a TBI and the complications that may arise from it also form a sizable portion of our chapter. Finally, steps will be discussed for cognitive rehabilitation and recovery from an injury of this magnitude.

No one willingly puts themselves in harm's way. The instinct for survival is too strong for that in most cases. There are very few professions in today's world that would involve harm to the body. Unsurprisingly, athletes and sports players are at the most risk of injuries themselves as part of their profession. One sport that has widely drawn the attention of neuropsychologists and neurologists is American football, or rugby as it is otherwise known. The 2015 film, *Concussion* documented the struggle of forensic pathologist Dr. Bennet Omalu against the National Football League in the United States after several football players were found to be suffering from chronic traumatic encephalopathy (CTE). It would not be fair to point the finger at just American football when similar cases were also found in the military and even among professional wrestlers.

A 28-year-old male was diagnosed with a closed head injury as a result of being in a coma and suffering from bilateral mydriasis. Some of the surgical procedures conducted were a ventriculostomy, in which the excess cerebrospinal fluid (CSF) is drained, and a hemicraniectomy, where a skull flap and the dura layer are removed in order to reduce the pressure on the brain. The patient had been in a motorcycle accident and had not been wearing a helmet

DOI: 10.4324/9781032640839-13

at the time. He had been found by the bystanders after the accident and had a Glasgow Coma Scale (GCS) score of 3 and presented with right parietal cephalohematoma and CSF otorrhea (the CSF was leaking through the ears). A computed tomography (CT) scan revealed that there was bruising to the left frontal and parietal regions, along with cerebral edema and a fractured skull. After the procedures were conducted, there was found to be an improvement in the GCS scores. After spending close to 21 days in the hospital, the patient was then discharged to a long-term care facility. A follow-up conducted three months later indicated external herniation (Nelson et al., 2016). The main objective of using a case study here is to highlight the unique terminology that comes into play when dealing with TBI cases. There are many of these, and while these might seem relevant to the field of neurosurgery and neurology, it is also crucial for aspiring clinical psychologists. Most terminologies, often gatekeeper information as these, would only be meaningful to the profession- als who study them. Being privy to them is a professional advantage as the severity of the injury to the brain would also dictate the level of deficits and the future course of action when it comes to rehabilitation.

Epidemiology of Traumatic Brain Injury

Understanding the rate of incidence and prevalence within the country is a prerequisite for adequate preparation for tackling the problem. In a paper on the epidemiology of TBI that was published at the beginning of the millen- nium, Gururaj (2002) highlighted the fact that there was a growing relation- ship between the growth of the country and the subsequent impact it has on social dynamics. Among the rise in the different types of injuries, TBI occupies a significant number due to its high impact on a developing nation like India. There are finite resources when it comes to dealing with TBI, and it has a more severe impact when compared to the same in developed nations. The leading cause of these seems to stem from the motor vehicle accidents that will keep increasing as more and more vehicles are introduced on Indian roads. In the city of Bengaluru, about 24% of injuries that were registered in the hospitals were cases of TBI. Extrapolating on this data would give an estimated 1.6 mil- lion people that sustain a TBI and have to get admitted to the hospital.

Consequently, about 200,000 of these also die from their injuries. These are all numbers that are based on estimates in the lower range. There is no national database for these numbers; hence these must be taken with caution. In a country that is transitioning from its villages to the cities, the rural-urban disparity also comes into the picture. In the last millennium, the most com- mon external cause of TBI seemed to be traffic accidents, particularly for men between the ages of 20 and 29. The second cause of injury sustained by falling seemed to be limited to children and older adults. Lastly, a few of the TBIs were also caused as a result of being assaulted in various situations, and the rest were due to other external causes as well. These are numbers that were published at the beginning of this millennium, and further surveys of this kind would yield a more accurate picture of the Indian scenario. Another recent study conducted by Shekhar et al. (2015) showed that males were more likely to be victims of TBI, and the leading cause for this was due to falling from heights. The second cause for the same was due to motor vehicle accidents.

The typical manifestation of TBI is headache, vomiting, along with loss of consciousness. Also commonly observed clinical presentation is the CSF otorrhea and ear and nasal bleeding. A systematic review of the quality of quantitative studies on the topic of TBI showed that there was a remarkable lack of quality in the methodology of the studies. It, however, also supported the epidemiology where road traffic accidents (RTAs), and falls were the most commonly reported cause of TBI. The studies conducted so far in the country are a good baseline for epidemiological studies that can inform future studies in this area (Massenburg et al., 2017). India has been lagging in terms of its commitment to combat most medical conditions. There are very lax road laws as well, and there is still resistance to wearing helmets despite repeated offenses. There is a severe need for changing this behavior as TBI has a poor prognosis in cases of road traffic accidents. It is the two-wheeler riders who are more likely to suffer the most, and it is this demographic that has to be explicitly targeted. It would require a combination of policy changes and traffic police coordination and cooperation. There have been many schemes and plans where the traffic police would also, at one point, provide helmets too, but there are no changes in helmet-wearing habits per se.

The rest of the world presents a similar situation when it comes to the causality of TBI. Falls and RTAs are the most reported causes, with falls being more commonly reported. A small difference lies in the fact that older people are more likely to experience TBI. The meta-analysis reported an incidence rate of 262 per 100,000 for TBI. Victims of TBI tend to be between the ages of 25 and 75, with a trend that shows that the elderly are likely to suffer more. Also, males are more likely to suffer from it than females. There is also a shift from RTAs to falls, which would be concurrent with the shift to the elderly (Peeters et al., 2015). Cases with severe TBI continue to come from RTAs, but there are more cases of mild traumatic brain injury (MTBI) than major ones. A study by Skandsen et al. (2018) in Norway focused on MTBIs showed that there was a need to update the epidemiological studies in the Nordic region. Most of the patients were male, and half of them were about 28 years old. They were mostly caused by falls followed by violence, sports, and RTAs. These causes were split along the ages, where the former was associated with the elderly and the latter with the younger groups. Another statistic that was unique to this study was that there were patients who were intoxicated at the time of admission. The fact that the median age of the patients was 28 seems to indicate that there is a level of carelessness that could be attributed to being young. Also, alcohol forms a part of the culture in Norway and is a risk factor for MTBI.

Mechanism of Impact

So far, we have been discussing the various types of causes for TBI. However, it is essential to understand that there are various severity of injuries that can occur. It is not limited to the surface of the skull and often impacts the brain at various levels. We shall discuss these in the following sections and elaborate on the types of deficits that these cause, along with the chances of recovery and rehabilitation. The impact on the head can cause damage at the neuronal level as well. The various parts of the neurons are different in terms of the damage that they can take and the amount of time it takes for them to recover as well. It can endure only a certain amount of stretching and tearing before it is completely lost. The strength of the axons to usually endure stress is affected as a result of such trauma. These microanatomic changes would mean that there are cognitive defects that originate at a cellular level. These are more difficult to heal as the neuron does not respond to healing at the microlevel. In cases of damage where the axon is not severed, there may be degeneration in the cell body, causing its death. It is known as retrograde degeneration, while when the damage causes the axon to degenerate, it leads to anterograde degeneration. In cases where there has been complete shearing of axons, there have been new axons sprouting. It is similar to rewiring areas after they have been damaged. There are also those areas that are more likely to be damaged due to excess connections. These connections are not well planned and would lead to behavioral disturbances as a result.

In order to understand this better, the classification of all TBIs can be divided into closed and penetrating head injuries. These are reasonably self-explanatory, where the penetrating head injuries indicate that a foreign object has lodged itself into the brain through the skull. These are harmful to the structure of the brain and lead to other associated complications like infections and hemorrhaging. In cases of closed head injuries (CHIs), two causes are common to all. These are acceleration and deceleration. The first case involves the brain being in a state of rest where it is acted on by an external force that pushes it into motion. Most of these cases are due to being hit in the head. The cases where deceleration is involved mean that the brain was traveling (in motion, not on vacation), and it came to a sudden halt. RTAs are typical scenarios for this type of injury. There are chances of the first impact injury where the brain hits the skull. The second time that the brain goes on to hit the skull again leads to a contrecoup injury (Zillmer & Spiers, 2001).

Nerve Injuries

When it comes to nerve injuries, they could be caused for various reasons. Seddon (1942) was one of the first who attempted to classify the different types of nerve injuries. It was during the 1863 US Civil War that the US Surgeon General established a unit that was to assist and treat nerve injuries. However, the surgeons and neurologists were sometimes mostly preoccupied with taking care of the soldiers rather than ascertaining the causes of the injuries. The Civil War was one of the bloodiest conflicts in the United States, and it led to many injuries as well. In the context of when Seddon classified nerve injuries, the world was in the midst of a Second World War. He noted that there were lesser nerve injuries in this one as compared to the US conflict, but this meant that these would be easier to classify, which would also assist in the clinical pathology of injuries of this nature. When diagnosing peripheral nerve injuries, the clinical manifestations tend to fall under two broad categories: loss of function and perversion of function. The former usually leads to paralysis of the affected areas and the loss of sensation like touch, temperature, and pain, among others.

The latter leads to muscular twitches and involuntary movement as well. There is also paresthesia (often described as a tingling sensation) along with the spontaneous sensation of pain. Three types of peripheral nerve injuries could be sustained as a result. These are known as neurotmesis, axonotmesis, and neuropraxia. Neurotmesis indicates the cutting of parts of the nerves and is a complete picture of a lesion. Neurotmesis usually consists of a neuroma which is a swelling at the site. There is both sensory and motor paralysis. In this case, there is some chance of recovery from motor paralysis but not sensory functions. Axonotmesis is a relatively less severe injury where there is damage to the fiber that leads to degeneration, but the myelin sheath and a majority of the cell body continue to remain active. Axonotmesis usually means that the fibers might be crushed or destroyed, but the underlying connective tissues remain active. These usually have a high chance of spontaneous regeneration. There is a severing of nerve fibers, but the supporting structure of the nerves is still active, which contributes to the regeneration process as well. Axonotmesis is functionally indistinguishable from neurotmesis, but the difference lies in the type of injury. Injuries from gunshots or blunt force trauma are more likely to lead to neurotmesis. Also, axonotmesis is more likely to heal quicker than neurotmesis, which would also provide a clue to the type of nerve injury.

Finally, neuropraxia is a mild nerve injury that typically leads to short-term paralysis that also recovers in a short period. It mostly impacts motor function and only partially affects sensory nerves. There is a perfect chance of recovery in this case as there is no peripheral degeneration either, as is observed in the other injuries. These shall now be looked at in much greater detail as well. The initial differentiation and etiology were elaborated on in another paper by Seddon (1942). It was an elaboration of his previous paper and led to detailed descriptions of the nerve injuries that he came to study. Before we continue with the details, it is crucial to note that the nature of the injuries and the complexity of the brain as an organ means that the types of real injuries that are usually sustained are a combination of the three types of injuries. Any combination of these could occur at any site of the brain. Another common syndrome when it comes to nerve injuries is Wallerian degeneration. Waller (1850) discovered this during his experiments on the nerves of a frog. The main objective was to observe whether there were any alterations to the nerve fibers after lesions. In order to not kill the animals, he only decided to cut one nerve of the pair. This decision gave him a clear indication of the healthy nerves vis-à-vis the lesioned nerves. There was considered loss of volume in the nerves which had been cut. The disorganization and atrophy of the nerves were seen as progressive and reduced parts of the medulla to the granular level as time went by. The main observation posited by Waller was the progressive nature of degeneration and the distal effect it had on the medulla. A few days of inactivity by the ligature of the glossopharyngeal nerve trunk led to disastrous effects on the medulla. This discovery drew attention to the effects of paralysis on nerve fibers and the role these fibers play in similar degeneration.

Neurotmesis

These are a complete anatomical division of the neurons and usually occur as a result of severe head injuries and penetrating head wounds. After the division, it was observed that there was regrowth until the point of division, but the development of scar tissue at the site was a barrier to spontaneous recovery from the peripheral degeneration. It also meant that the two sides were not close enough to reach and reform the connection again. In terms of motor function, all kinds of paralysis and degeneration of muscles tend to happen below the site of the lesion. Muscles tend to atrophy when there is a lack of stimulation which is observable in the second-week post-injury. There was also sudomotor (sweat glands), vasomotor (constriction of blood vessels), and pilomotor (hair reflex in response to excitement, goosebumps) paralysis. Wallerian degeneration occurs in these cases as well. In terms of the senses, there is a loss of touch, temperature, and pain. It might also have distal effects where if the injured site happens to be the arms, there might be loss of sensation in the fingers. While these effects are long term, the area they affect is generally smaller than expected. It could be attributed to the body adapting and, over time, other nerves taking over the functions of the damaged nerves and compensating for the deficit. While there are cases where the regeneration process completes itself, it is rare, and it is almost always imperfect.

Axonotmesis

As discussed earlier, one of the ways that axonotmesis differs from neurotmesis is by the type of injury. In this case, closed fractures and dislocations are more likely to cause axonotmesis. The effects of the injury are indistinguishable from neurotmesis. The main difference is that the recovery is spontaneous. It is slow, but not too slow, as in cases of repaired nerves. The regeneration starts from the closest areas and moves to the distal locations. The muscular atrophy that is present in neurotmesis is absent in axonotmesis. The muscles recover fairly quickly themselves and often lead to a balanced growth of volume and functionality. The sensory functions also recover quickly, and this is the case only after axonotmesis. The recovery is not as perfect in any other case.

Animal experimentation showed that when the nerves were compressed using forceps, the axon tended to die out and degenerate. Wallerian degeneration was also present in this case. However, when the myelin sheath was intact and in the absence of any scar tissues, wounds, and blood, it was a simple matter of the nerves reforming in the degraded area. Those, as mentioned earlier, are usually in the way of spontaneous recovery when it comes to neurotmesis. The open wounds that usually accompany neurotmesis are a hotbed of infections and altered intraneural topography. In cases where nerves have been repaired, there is a connection and mismatch on the two sides, but the recovery in axonotmesis is natural such that the nerves tend to follow their old paths quickly and efficiently. There is no confusion with regard to where one fits the other. It provides proper diagnostic criteria for a surgeon assessing neural injuries.

Neurapraxia

These are typically short-term paralysis that leads to an interruption in conduction within the nerves. Some of the characteristics of neuropraxia are that motor paralysis is higher than sensory paralysis. However, there is little muscular atrophy involved in these cases. The motor recovery is almost instantaneous, but it does not follow any set pattern. The nature of the injury that causes neurapraxia is varied between instantaneous, prolonged, and intermittent injuries. Examples of an instantaneous injury would be gunshot wounds and fractures.

Meanwhile, prolonged cases were due to compression of muscles and nerves due to bandages and being stuck under debris. Other causes were also due to airplane travel, where the body is forced to maintain an uncomfortable position for a prolonged time. A case of neurapraxia was also observed in a case where the patient had been using crutches. In terms of neuropraxia as a clinical syndrome, the level of paralysis was indistinguishable from neurotmesis. The motor function was affected, but it recovered soon. There was no Wallerian degeneration in these cases. It could be because recovery was reasonably quick. There were subjective sensory experiences associated with the various cases of neurapraxia. There was a sensation of numbness or tingling, which is also often described as the pins and needles feeling. Nevertheless, most did not have any significant complaints concerning their sensation.

Assessing the Severity of Brain Injury

When it comes to the assessment of the severity of brain injury, some tools are used in the field, and other tools are preferred by researchers. This distinction is vital as most cases of head injury tend to be in the emergency sections of hospitals. The most commonly used scale in this regard is the Glasgow Coma Scale (GCS) (Teasdale & Jennett, 1974). This scale continues to be relevant 50 years after its inception that stands as a tall testimony to its utility and rigor. It gives the trauma team a quick assessment of the depth of coma or unconsciousness by a quick checklist of various symptoms associated with the state. The various symptoms cover the domains of language, consciousness, and motor. The items on the GCS are mainly about the eyes, whether the patient is well oriented and can respond to questions, and finally, whether they can perform a simple motor task in response to a command.

Scores higher than 13 might indicate mild confusion, whereas scores less than five are indicative of deep coma. It also remains a good measure of prognosis and outcome. Scores higher than eight would imply good chances of recovery, whereas anything lower than seven would mean higher chances of mortality. The main indication of the severity of the injury also stems from the induction of coma. Coma is complicated in that it does occur all the time. More often, any injuries to the reticular activating system (RAS) would typically induce coma as well. Coma lies on a continuum; however, it can still be assessed as one would the various levels of sleep. GCS is a tool that can help with that. However, there are also limitations to the tool, as it relies on observations by the clinician. Also, the use of a coma as an indicator of brain damage is not a good indicator. As discussed earlier, any damage to the RAS or the parts of the brain associated with consciousness might bring about a coma, even if the rest of the brain is relatively damage free. The scores on the GCS are also a good indicator of the severity of the neuropsychological deficits that might have manifested as a result of the injuries. It sets up the road to recovery and rehabilitation as well. Other factors that might indicate the chances of a functional recovery would be the number of days the patient was in a comatose state. Some of the other areas that need to be assessed as part of this would be checking the airway, cervical spine protection, breathing, and hemorrhage control. Other standard practices are that oxygen must be given so that hypoxemia does not occur. In terms of imaging techniques, a CT scan or a positron emission tomography (PET) scan would reveal more that is not possible through a cursory observation (Liew et al., 2017).

Complications from Traumatic Brain Injury

The effects of head injuries are not restricted to the brain but are more likely to cause other complications and effects on other areas. These are in the form of edema, hernias, hemorrhages, epilepsy, and loss of orientation or confusion. It would be a broad look at what changes happen apart from the nerve injuries that we have discussed so far.

Edema

Most parts of the body are likely to swell if they come under trauma from an external force. Similarly, the brain would also swell in case of an injury, but the brain does not have any space to swell. The skull will not yield to any pressure from such swelling, and this usually leads. Nevertheless, there are spaces, as discussed, between the different layers. There are intracranial pressure (ICP) buildup points that can lead to death in other cases as well. These are primarily associated with cerebral edema that restricts the flow of blood supply to the brain. Some of the most common symptoms of cerebral edema may include headaches, nausea, vomiting, difficulty speaking, and loss of consciousness. These are all ways the body tries to naturally relieve the pressure due to the swelling in the brain. Medications and, in more severe cases of edema, surgery is done in order to relieve the pressure on the brain (Kandola, 2018).

Brain Herniation

These are displacements or deformation of different parts of the brain. These could be due to a variety of causes, like hemorrhages and infections. These also consequently lead to high intracranial pressure. A particular form of herniation is transtentorial herniation which is the downward movement of the parahippocampal gyrus and the uncus of one or both temporal lobes and the tentorial hiatus. It could give rise to pressure on cranial nerves as well as effectively cutting them off and leading to nerve injuries, as previously described. Edemas are also at risk of giving rise to hernias which is why the primary focus of any head trauma team is on reducing the ICP in these cases.

Extradural and Subdural Hemorrhage

The dura mater is one of the meninges that form the layers of protection between the skull and the brain. In case of any injury, the cerebral blood vessels might rupture, leading to internal bleeding between the spaces around the dura mater. The dura mater is a secure layer that encompasses the brain and initially leads to suppression of the bleeding. However, it can eventually generate too much downward pressure, leading to brain herniation. The subdural hematoma usually occurs in between the dura and the arachnoid spaces. These are variable in their development and depend on the rate of bleeding as well. The extradural hematoma, as indicated by the name, occurs in the layer between the skull and the dura. The large meningeal artery might get damaged and lead to a hematoma. Most of the hematomas are usually treated in cases where the meningeal arteries rupture, leading to massive internal bleeding. In these cases, the most common procedure is to drill a burr hole in the skull and place a shunt to drain the excess blood within the brain. It is the quickest method, and timely diagnosis is necessary for it.

Intracranial Bleeding

In cases where there are clots, there are also chances of microscopic hemorrhages that can be formed within the cranium whenever the white matter gets affected. These usually also appear as focal lesions that look like intracranial hematomas. The previous section would not count as intracranial; they appear outside the brain. These are difficult to treat and, in most cases, require emergency surgery.

Post-Traumatic Epilepsy

It is a relatively unknown entity. In most cases, there is no plausible reason as to why epilepsy occurs. These are paroxysmal electrical discharges that are similar to the electrical circuits in households blowing a fuse due to overcharging. (Verellen & Cavazos, 2010) provide an overview of post-traumatic epilepsy, which is provoked by a TBI. Not all cases of TBI are likely to lead to post-traumatic epilepsy (PTE). Some of the predictors that are an estimate of PTE are the severity of the TBI itself. The previous complications are a prerequisite for PTE as well. The type of lesion, along with its location and size, are also good indicators of PTE. Scar tissues that are usually formed in cases of neurotmesis and alterations in the neural membrane function also lay the groundwork for seizures. The injury itself is secondary, while the lesions caused by them are the main reason for the seizures. Coma could also induce seizures (Zillmer et al., 2007).

Post-Traumatic Confusion

There might be a loss of consciousness; it might be a temporary thing, but there might also be cases where it is more permanent and enduring. There need to be more articles on these kinds of disorders. There might also be more chances that these are some things that do not even make it to the hospitals due to their transient nature. These are some of the more commonly used terms in neurology concerning brain trauma and injury. Sherer et al. (2008) showed that the effects of post-traumatic confusion post-TBI had poorer outcomes in terms of employability and productivity; the more severe the confusion. Most patients who are recovering from TBI have brief periods of disorientation and are unable to form any new memories. There have been recent proposals to replace the term post-traumatic amnesia with post-traumatic confusion. There was also the presence of psychotic-type symptoms, which also affected productivity post a year after the injury.

In the next sections, we will be dealing with the effects that these have on the overall development and functionality of the brain. Our discussion until this point has been substantial with the biology of the entity. It was done to build confidence and knowledge about the types of injuries and the damage these cause to the brain itself. In the future, these are also related to the deficits that we see in the next section.

Neuropsychological Manifestation of Traumatic Brain Injury

When it comes to the effects of TBI on the brain, there are more likely to be other psychiatric disorders like major depression, bipolar disorder, generalized anxiety disorder, panic disorder, and post-traumatic stress disorder (PTSD). The neuropsychiatric effects of TBI which is influenced by the severity and type of injury, past psychiatric diagnosis, childhood problems, if any, and finally, social support along with preexisting substance abuse problems. Most of the disorders are also related to the type of injury that was sustained due to the TBI. The age of the patient at which the injury was sustained was also a significant factor. It can be attributed to the fact that there is more likely to be considered cognitive resources and plasticity potential in the younger generation than in the older patients. Older patients are more likely to suffer cognitive decline and more significant time to recover, along with more extended periods of agitation (Ahmed et al., 2017). For the sake of simplicity, it would be easier to divide the following discussion along the lines of cognitive deficits and neuropsychiatric effects of TBI.

Cognitive Deficits

These are the deficits that typically include impairment of arousal, attention, concentration, and executive function, among other things. There has been a proposal that cognitive decline can be divided into four categories based on the phases of the TBI. The first phase is the loss of consciousness that usually occurs immediately after the injury. The second phase is when the patient has regained consciousness, and there is a combination of behavioral and cognitive issues. In this phase, the patient is unable to recall events, and the sequence of time is lost as well. These two phases themselves are more likely to last anywhere from a few days to a month. The third phase, immediately after this, is when there is a 6–12-month period of recovery where the cognitive functions are recovered. The last phase is when the permanent effects of the injury sets in. There are problems with information-processing speed and attention along with short- and long-term memory deficits as well. There are also issues with executive functions and mental inflexibility. It has also been described as akin to dementia as a result of head trauma (Rao & Lyketsos, 2000). Warden et al. (2006) described the levels of cognitive domains. These are the most common divisions under cognitive functions. These will be briefly discussed individually now.

Attention

Impairment of attention is present in all levels of severity when it comes to TBI. Some of the patients have reported that there was a loss of train of thought and that there was frequent zoning out as well. There are also derived mechanisms that get affected easily. The diffused damage that is a part of TBI impacts the ability to attend or concentrate. Other abilities that are related to this impairment are increased distractibility and difficulty in working on more than one task. Even simple tasks end up taking more time than usual.

Memory

These are among the most common complaints when it comes to cognitive complaints in people with TBI. Memory is a domain closely connected to attention and executive functioning. These may affect the way that encoding and retrieval take place in the brain. Episodic memory is among the worst affected in these cases. There are also effects on prospective memory where the patients are not able to pay their bills on time or any other types of tasks. There have been cases of post-traumatic amnesia as well.

Executive Functions

The prefrontal brain systems are particularly vulnerable to diffuse and focal damage. Verbal and design fluency is impaired, and conceptual reasoning and flexibility would also be affected. Categorization and response strategies are complicated, as well.

Perception

The occipital lobe is located in the back of the head and therefore is nearly as exposed as the frontal lobe. It is more likely to get affected as a result. Visual dysfunction predominates as part of a general cognitive decline as part of the deficits due to TBI due to this vulnerable position.

Language

Anomia and word-finding difficulty are frequent in these cases. Aphasias are also common in language deficits due to TBI. The presence of aphasias is indicative of more severe cognitive decline as well.

Intelligence

Both the performance and verbal IQ are affected in these cases. However, verbal IQ is quick to recover as compared to performance IQ, which might be impaired as far as three years (Kaur, 2011).

Neuropsychiatric Deficits

Some of the major risk factors for the development of neuropsychiatric disorders after TBI are older age, arteriosclerosis, and alcoholism. Some of the other risk factors are poor interpersonal relationships, financial instability, and marital discord, which might act as barriers to proper recovery from the disorder. The disorders that come across from TBI are difficult to describe. The injury that occurs more often in these cases is broadly classified as frontal lobe syndrome, temporal lobe syndrome, aggressive disorders, and personality changes. However, these are not entirely accurate, and therefore, it would be prudent to reclassify these according to the symptoms and the phenomenon. These would be mood disorders, anxiety disorders, psychosis, apathy, and behavior disorders. Mood disorders are a result of the disruption of biogenic amine-containing neurons through the basal ganglia and the frontal-subcortical white matter. The lesions in the left dorsolateral frontal and left basal ganglia would also increase the chances of major depression. The anxiety that patients with TBI experience are often a manifestation of a generalized anxiety disorder (Rao & Lyketsos, 2000). There are likely to be changes in the functionality of neurotransmitters post-TBI that usually lead to psychiatric manifestations. Catecholamines are carried by monoaminergic projections, which, when damaged, affect connected pathways. There is also a decrease in dopamine levels that leads to poor prognosis. Damage to areas within the brain usually leads to problems in the functions that those areas play. Depression, in these cases, is as much a result of biology as it is psychology.

There is an extreme reduction in the functionality of patients with TBI. Hence, the depression that manifests as a result is not the typical case, but this is accompanied by apathy and impaired thought processes. The head injury is severe enough that the patient experiences post-traumatic delirium as well. It is characterized by restlessness, disorientation, and in more extreme cases, hallucinations and delusions (Ahmed et al., 2017). Depression has become a frequent follower of TBI. There are also difficulties in the differentiation of manifestation of depressive symptoms that come up as part of adjustment or due to depression itself. The boundaries between the different types of symptoms are blurred. Other disorders like Obsessive-compulsive disorder, psychotic disorders, and mania are uncommon. The manifestation of mania is more likely to be in the form of aggression and irritability rather than euphoria.

On the other hand, it has now been established that MTBI and PTSD might be mutually exclusive, particularly in fact, due to the circumstances under which TBI tends to occur. Similarly, the use of alcohol has been one of the causes of TBI. They are more likely to lead to more issues with substance abuse, as well. Apathy, affective lability, aggression, and some symptoms of paranoia were also among a few personality changes that were observed as part of personality changes post-TBI. No known pathophysiology directly connects TBI to psychiatric disorders, while psychosocial aspects seem to predominate depression post-TBI (Schwarzbold et al., 2008).

Types of Rehabilitation

As with most rehabilitation cases, there are likely to be two modes of rehabilitation. The traditional method has always been one where there is an emphasis on manual or offline exercises. With the advent of affordable technological advancements, there are more and more online and digital methods of rehabilitation and exercise. Cicerone et al. (2011) believed that when it comes to rehabilitation for TBI, it would be valid if attention training and metacognition were used. Internal strategies and external devices were the way forward in cases where there were mild memory impairments. Cognitive rehabilitation involves the usual batteries of neuropsychological assessment and identification of specific behaviors that would need to be changed. Evaluation of these remediation processes follows frequent changes as well in most cases (Kaur, 2011).

For cases where the patient has attention problems, the use of the Attention Process Training module (Sohlberg & Mateer, 1987) is commonly cited. It is a hierarchical, multilevel program that was specifically designed to address the issues of attention and concentration in a person with brain injury. It is a cognitive remediation training program that follows five to ten weeks of specific attention training. It advocated the use of tests such as backward digit span, serial number sets, and backward spelling. It elaborates on five levels of attention, namely, focused, sustained, selective, alternating, and divided. These levels also informed the types of tasks that were developed for the model, like detecting auditory number targets, responding to a string of stimuli, the same task being presented with background noise for selective attention, addition and subtraction, and finally, simultaneous multiple attention tasks. Wherever possible, these were conducted using computers so that it was easier to record and score the progress of the patients.

When it comes to memory, the same techniques that were discussed in the chapter of memory are used for rehabilitation when it comes to memory deficits in patients with TBI. The only difference is that there is lesser reliance on visual imagery as these are difficult after brain injuries. So, popular methods like the method of loci would not be used for patients with TBI. Nevertheless, this could be used for people that have suffered verbal deficits rather than visual deficits. Instead, there would be more usage of external aids like mobile phones or alarm clocks that would be useful for cues. Language skills are easily trained by the use of word search games and crossword puzzles. Finally, visuoperceptual functions are impaired, which can be rectified by having the patients draw figures, do facial recognition, do puzzle assembly, and do mazes. Lastly, when there are executive function deficits, there are more chances of recovery when routines, exercises, and activities are taught to the patients. Teaching the patients the use of planners and the like are also a suitable method for utilizing their deficit function of planning and goal setting.

Computer-Assisted Cognitive Rehabilitation in Traumatic Brain Injury

The nature of the client's deficits and their computer literacy are vital factors in the usage of computer-assisted rehabilitation. These are also crucial factors when it comes to using computers as part of vocational training as well. It usually leads to positive responses from the patients. There was an effect of novelty when it was first presented. In particular, skill sets such as cognitive endurance, eye-hand coordination, visual tracking/sequencing, and problem-solving skills are more comfortable with assessing and rehabilitating on a computer (Kaur, 2011). Fetta et al. (2017) conducted a review of these technological rehabilitation techniques and showed that there was no substantial evidence for computer-based interventions for TBI. There are no robust research designs to suggest the utility of using computer-based rehabilitation techniques. These are good for improving memory functions but little for other domains (Bergquist et al., 2009). Most studies also included standard rehabilitation techniques that did not specify their level of physical activity, leading to a confound. A computer-assisted cognitive rehabilitation (CBCR) approach taps into the brain's neuroplasticity as a method apart from conventional techniques. It restores partial functionality of cognitive domains and other aspects of the brain. These are readily available as well as easily modifiable to suit individual needs. Nevertheless, the evidence remains controversial, and studies that show evidence lack proper research designs and are not replicable (Li et al., 2021).

Research Perspective

In recent decades researchers have made tremendous progress in TBIs research. The advancement in neuroimaging, neuromodulation techniques for the assessment and intervention of brain activity, application of virtual reality, and neurofeedback as intervention tools have been major contributors to the progress. Scientists have also been able to build a more profound understanding of brain functions post-TBI studies have shown the association of TBI with various psychiatric and neuropsychological disorders, which is helpful while building a comprehensive intervention plan. The advancement of methods is also making it possible to study patients from different age groups, including children and older populations.

One of the most prevalent causes of cases of TBIs is injury from sports and related activities, sports-related concussions being the most common injury among all others. As we have understood from the readings in the chapter, post-TBI care needs to be multifaceted, as there can be symptoms that are often sidelined but might indicate underlying pathophysiology. It is very much necessary to develop a comprehensive, evidence-based rehabilitation program for TBIs in such scenarios to address all the clinical conditions post-TBI. Though these injuries have been very common, literature addressing evidence-based clinical recommendations is still scarce (Marklund et al., 2019). In their review of available methods of assessment and rehabilitation, they identified the need to incorporate advanced technologies in cognitive rehabilitation as a part of the rehabilitation program post-TBI. In a recent review by Howlett et al. (2022), authors presented that dysfunction post-TBIs includes not only physical issues but also disrupted mental health, cognitive dysfunction, and emotional irritability.

Looking at advancement in TBI in different age groups, it can be seen that TBI causes many trauma deaths and disability in infants and children (Dewan et al., 2016). Since impact of brain injuries can lead to varying and more severe impacts in children as compared to adults, it's very important to have accurate and faster diagnosis procedures for better management of after effects (Araki et al., 2017). Biomarkers in the pediatric population have significant potential applications ranging from faster diagnosis in emergency care and intensive care unit, assessment of TBI severity and prognosis, and future facilitating therapeutic advancement in clinical trials (Wang et al., 2018). A systematic review of biomarker studies in the pediatric population has revealed serum biomarkers as the most studied biomarkers (Marzano et al., 2022). Studies indicate that S100B and NSE are the most intensively investigated biomarkers, UCH-L1 and GFAP being the subsequent ones. Some of the future directions for biomarker studies are exploring CSF biomarkers, neuroimaging biomarkers, and a larger sample pool with a specific population within the pediatric age group to acquire more specific information.

Neuroimaging and brain-stimulating tools are also being tested for their use in assessment, therapeutic advancement, and intervention plans. Studies have also analyzed the uses of non-invasive brain stimulation techniques in the rehabilitation of cognitive dysfunction in TBIs (Ahorsu et al., 2021). Non-invasive brain stimulation techniques include transcranial direct current stimulation and transcranial magnetic stimulation. These techniques are found to be moderately effective, however, they still need more prospective cohort studies. Further studies need to be done with a powered sample size, efficiently considering the confounding variables, and effective experiment design considering every participant's neurophysiopathology. Studies have indicated that repetitive transcranial magnetic stimulation (rTMS) has positive effects in treating some of the post-TBIs' mental health symptoms, including depression, visual neglect, etc. (Pink et al., 2021). However, research needs to be done to have a better understanding of the guidelines for using rTMS and its longitudinal effects on different populations. Scientists have also explored the uses of magnetic resonance imaging (MRI) in investigating neuropathology in mild TBIs. MRI proves to be one of the very effective imaging methods in studying structural changes due to neuropsychological conditions associated with traumatic head injury. In the future, researchers need to look into a multimodality approach instead of just using a volumetric approach while studying the brain post-TBIs to gain more valuable and reliable insights (Bigler, 2023). Another emerging area of research in cognitive rehabilitation post-TBI is the use of virtual reality. Since virtual reality has uses in both motor and cognitive rehabilitation, it has the potential to be used as a multifactor rehabilitation tool.

There is a need to investigate the diversity within the TBI population. Cohort studies are needed to identify the factors contributing to the mental health sequelae post-TBI depending upon the nature of the injury, severity level, and clinical and demographic variability. This will help in designing a post-injury management plan addressing the specific needs and challenges faced by different subpopulations in a more effective way. The effect of the recurrence of TBIs on mental health and recovery is also one of the understudied topics. It needs longitudinal studies for the development of preventive measures and to mitigate the potential long-term effects.

The vast literature on TBI clearly indicates that there is a variation in post–TBIs clinical and physiological symptom presentation which makes it crucial for those who are dealing with TBI to consider the individualistic nature of post-TBI challenges. It is also equally important to pay attention to specific medical complications associated with TBI, such as paroxysmal sympathetic hyperactivity, post-traumatic hydrocephalus, and neuroendocrine dysfunctions, and tailor neurorehabilitation approaches accordingly (Oberholzer & Müri, 2019).

References

Ahmed, S., Venigalla, H., Mekala, H. M., Dar, S., Hassan, M., & Ayub, S. (2017). Traumatic brain injury and neuropsychiatric complications. *Indian Journal of Psychological Medicine*, *39*(2), 114–121.

Ahorsu, D. K., Adjaottor, E. S., & Lam, B. Y. H. (2021). Intervention effect of non-invasive brain stimulation on cognitive functions among people with traumatic brain injury: A systematic review and meta-analysis. *Brain Sciences*, *11*(7), Article 7. https://doi.org/10.3390/brainsci11070840

Araki, T., Yokota, H., & Morita, A. (2017). Pediatric traumatic brain injury: Characteristic features, diagnosis, and management. *Neurologia Medico-Chirurgica*, *57*(2), 82–93. https://doi.org/10.2176/nmc.ra.2016-0191

Bergquist, T., Gehl, C., Mandrekar, J., Lepore, S., Hanna, S., Osten, A., & Beaulieu, W. (2009). The effect of internet-based cognitive rehabilitation in persons with memory impairments after severe traumatic brain injury. *Brain Injury*, *23*(10), 790–799.

Bigler, E. D. (2023). Volumetric MRI findings in mild traumatic brain injury (mTBI) and neuropsychological outcome. *Neuropsychology Review*, *33*(1), 5–41. https://doi.org/10.1007/s11065-020-09474-0

Cicerone, K. D., Langenbahn, D. M., Braden, C., Malec, J. F., Kalmar, K., Fraas, M., Felicetti, T., Laatsch, L., Harley, J. P., & Bergquist, T. (2011). Evidence-based cognitive rehabilitation: Updated review of the literature from 2003 through 2008. *Archives of Physical Medicine and Rehabilitation*, *92*(4), 519–530.

Dewan, M. C., Mummareddy, N., Wellons, J. C., & Bonfield, C. M. (2016). Epidemiology of global pediatric traumatic brain injury: Qualitative review. *World Neurosurgery*, *91*, 497–509.e1. https://doi.org/10.1016/j.wneu.2016.03.045

Fetta, J., Starkweather, A., & Gill, J. M. (2017). Computer-based cognitive rehabilitation interventions for traumatic brain injury: A critical review of the literature. *Journal of Neuroscience Nursing*, *49*(4), 235–240. https://doi.org/10.1097/JNN.0000000000000298

Gururaj, G. (2002). Epidemiology of traumatic brain injuries: Indian scenario. *Neurological Research*, *24*(1), 24–28.

Howlett, J. R., Nelson, L. D., & Stein, M. B. (2022). Mental health consequences of traumatic brain injury. *Biological Psychiatry*, *91*(5), 413–420. https://doi.org/10.1016/j.biopsych.2021.09.024

Kandola. (2018). *Cerebral edema: Symptoms, causes, treatment, outlook.* https://www.medicalnewstoday.com/articles/322475#symptoms

Kaur, L. (2011). Neuropsychological impairments and remediation in traumatic brain injury. In V. Sharma & S. Malhotra (Eds.), *Clinical neuropsychology* (pp. 151–171). Harprasad Institute of Behavioral Sciences.

Li, Y., Zhang, S., Snyder, M. P., & Meador, K. J. (2021). Precision medicine in women with epilepsy: The challenge, systematic review, and future direction. *Epilepsy & Behavior*, *118*, 107928. https://doi.org/10.1016/j.yebeh.2021.107928

Liew, B., Zainab, K., Cecilia, A., Zarina, Y., & Clement, T. (2017). Early management of head injury in adults in primary care. *Malaysian Family Physician: The Official Journal of the Academy of Family Physicians of Malaysia, 12*(1), 22.

Marklund, N., Bellander, B.-M., Godbolt, A. K., Levin, H., McCrory, P., & Thelin, E. P. (2019). Treatments and rehabilitation in the acute and chronic state of traumatic brain injury. *Journal of Internal Medicine, 285*(6), 608–623. https://doi.org/10.1111/joim.12900

Marzano, L. A. S., Batista, J. P. T., de Abreu Arruda, M., de Freitas Cardoso, M. G., de Barros, J. L. V. M., Moreira, J. M., Liu, P. M. F., Teixeira, A. L., Simoes e Silva, A. C., & de Miranda, A. S. (2022). Traumatic brain injury biomarkers in pediatric patients: A systematic review. *Neurosurgical Review, 45*(1), 167–197.

Massenburg, B. B., Veetil, D. K., Raykar, N. P., Agrawal, A., Roy, N., & Gerdin, M. (2017). A systematic review of quantitative research on traumatic brain injury in India. *Neurology India, 65*(2), 305.

Nelson, C. G., Elta, T., Bannister, J., Dzandu, J., Mangram, A., & Zach, V. (2016). Severe traumatic brain injury: A case report. *The American Journal of Case Reports, 17,* 186.

Oberholzer, M., & Müri, R. M. (2019). Neurorehabilitation of traumatic brain injury (TBI): A clinical review. *Medical Sciences, 7*(3), 47.

Peeters, W., van den Brande, R., Polinder, S., Brazinova, A., Steyerberg, E. W., Lingsma, H. F., & Maas, A. I. (2015). Epidemiology of traumatic brain injury in Europe. *Acta Neurochirurgica, 157,* 1683–1696.

Pink, A. E., Williams, C., Alderman, N., & Stoffels, M. (2021). The use of repetitive transcranial magnetic stimulation (rTMS) following traumatic brain injury (TBI): A scoping review. *Neuropsychological Rehabilitation, 31*(3), 479–505.

Rao, V., & Lyketsos, C. (2000). Neuropsychiatric sequelae of traumatic brain injury. *Psychosomatics, 41*(2), 95–103.

Schwarzbold, M., Diaz, A., Martins, E. T., Rufino, A., Amante, L. N., Thais, M. E., Quevedo, J., Hohl, A., Linhares, M. N., & Walz, R. (2008). Psychiatric disorders and traumatic brain injury. *Neuropsychiatric Disease and Treatment, 4*(4), 797–816.

Seddon, H. (1942). A classification of nerve injuries. *British Medical Journal, 2*(4260), 237.

Shekhar, C., Gupta, L., Premsagar, I., Sinha, M., & Kishore, J. (2015). An epidemiological study of traumatic brain injury cases in a trauma centre of New Delhi (India). *Journal of Emergencies, Trauma, and Shock, 8*(3), 131. https://doi.org/10.4103/0974-2700.160700

Sherer, M., Yablon, S. A., Nakase-Richardson, R., & Nick, T. G. (2008). Effect of severity of post-traumatic confusion and its constituent symptoms on outcome after traumatic brain injury. *Archives of Physical Medicine and Rehabilitation, 89*(1), 42–47.

Skandsen, T., Einarsen, C. E., Normann, I., Bjøralt, S., Karlsen, R. H., McDonagh, D., Nilsen, T. L., Akslen, A. N., Håberg, A. K., & Vik, A. (2018). The epidemiology of mild traumatic brain injury: The Trondheim MTBI follow-up study. *Scandinavian Journal of Trauma, Resuscitation and Emergency Medicine, 26*(1), 1–9.

Sohlberg, M. M., & Mateer, C. A. (1987). Effectiveness of an attention-training program. *Journal of Clinical and Experimental Neuropsychology, 9*(2), 117–130.

Teasdale, G., & Jennett, B. (1974). Assessment of coma and impaired consciousness: A practical scale. *The Lancet, 304*(7872), 81–84.

Verellen, R. M., & Cavazos, J. E. (2010). Post-traumatic epilepsy: An overview. *Therapy*, *7*(5), 527.

Waller, A. V. (1850). XX. Experiments on the section of the glossopharyngeal and hypoglossal nerves of the frog, and observations of the alterations produced thereby in the structure of their primitive fibres. *Philosophical Transactions of the Royal Society of London*, *140*, 423–429.

Wang, K. K., Yang, Z., Zhu, T., Shi, Y., Rubenstein, R., Tyndall, J. A., & Manley, G. T. (2018). An update on diagnostic and prognostic biomarkers for traumatic brain injury. *Expert Review of Molecular Diagnostics*, *18*(2), 165–180. https://doi.org/1 0.1080/14737159.2018.1428089

Warden, D. L., Gordon, B., McAllister, T. W., Silver, J. M., Barth, J. T., Bruns, J., Drake, A., Gentry, T., Jagoda, A., & Katz, D. I. (2006). Guidelines for the pharmacologic treatment of neurobehavioral sequelae of traumatic brain injury. *Journal of Neurotrauma*, *23*(10), 1468–1501.

Zillmer, E. A., & Spiers, M. V. (2001). *Principles of neuropsychology*. Wadsworth/ Thomson Learning.

11 Neurodevelopmental Disorders of Childhood

Introduction

There are two ages that human beings are at their most vulnerable. The first is when they are born and live as children, and the second is when they are old and only a few years from death. Childhood development holds the key to the future that can either be the best or worst form of living. As the child develops inside the womb, the maintenance of an atmosphere that is conducive to its proper growth and development is a delicate balance. The main factors that might play a role in that phase are genes, chromosomes, and environmental impacts. The development of a neurodevelopmental disorder means that these forces play a part in the early stages of brain development. However, their effects persist through much of the individual's life. The first topic of discussion is the weakness of the brain at its developing stages and how this might be different from an adult's brain. These are important as in evaluating them, we might be able to predict the future course of the problem and how to tackle it. Some of the disorders that are reviewed are hydrocephalus (HC), Williams syndrome (WS), Turner's syndrome (TS), and fetal alcohol syndrome (FAS). Finally, the prevalence and manifestations of these disorders, along with their pathology and assessment techniques, are also discussed. The techniques for rehabilitation are also explored toward the end of the chapter.

Neurodevelopmental disorders are a group that typically occurs in children before they start going to school – impairments in brain development impact other functions like learning, memory, emotion, language, and motor control. Biology plays a large part in the etiology, and this often manifests itself in the form of developmental milestones. This group contains disorders specific to a particular function, like specific learning disorders, and widespread problems like social and intellectual functioning. Some other disorders under discussion, along with the ones above, are autism spectrum disorder, attention-deficit/hyperactivity disorder (ADHD), and Tourette's (American Psychiatric Association, 2013).

Vulnerability and Plasticity of the Developing Brain

Due to various structural and functional anomalies in the genes and chromosomes, along with environmental factors, the brain might not be able to

DOI: 10.4324/9781032640839-14

develop correctly. Brain disorder like phenylketonuria (PKU) can either be inherited or results from alterations of the cerebrum. Some of the other environmental factors that upset the delicate balance of the child's environment are alcohol consumption during pregnancy. Furthermore, substances like teratogens could be present or introduced at various stages of fetal development that can cause long-term central nervous system deficits. These could be classified as agenesis, the failure of an organ to develop. Alternatively, dysgenesis indicates the abnormal development of an organ. The effects teratogens can have on the brain and the body can be concentrated or diffused. Common examples of teratogens are widespread diseases (such as influenza and mumps), sexually transmitted diseases (like syphilis and AIDS), drugs (cocaine, alcohol), harmful chemicals (mercury, carbon monoxide), and radiation (from potential sources like X-rays and radioactive materials). Weiss (2000) showed that children were vulnerable due to the presence of various neurotoxic hazards. Historically speaking, the entire field of teratology was born from the need to study the effects of thalidomide. It was a drug that was prescribed as a sedative to pregnant women to relieve nausea. However, it also led to malformation of limbs and other congenital disabilities in the babies.

Mercury, as a chemical, has now been attributed to causing pink disease (acrodynia). It is a condition of pain and pink coloration of the hands and feet in response to chronic exposure to heavy metals such as mercury. It was in the year 1947 that this link was established. Mercury was used in the form of calomel, which was a component in teething powders. Once this was removed from the powders, pink disease rates plummeted. These incidents have ensured that chemicals are being tested for their effects on the developmental milestones of babies. The most common source tends to be FAS. At its worst, it causes craniofacial dysmorphologies, and other effects follow lower levels of alcohol intake.

The basis for the development and formation of most structures present in the brain can be attributed to stem cells. These are numerous in the early stages and are the proverbial clay that can be shaped to make anything possible. As these are utilized, various structures form parts of systems in the brain. However, this also means that the brain at this stage is malleable and can be susceptible to changes due to various factors. These are a few of the many ways that the brain of a neonate differs from the brain of a full-grown adult.

Child and Adult Brain: Various Structural and Functional Differences

There are some rather obvious ways in which the adult brain is different from the children. The adult brain is more advanced and mature. What this also means is that lesions affect these two in various manners. Different mechanisms govern functions. In the child's brain, the methods of acquiring mental capabilities are impaired, while the adult loses the capabilities that they have acquired till that point. The mature brain is more likely to be behaviorally stable and, therefore, predictable. The child's brain can develop cognitively and behaviorally in leaps and bounds. There are subtle variations that are influenced by the current developmental level of the child as well as their social, psychological, and physical environment. Coming back to lesions and their effect, the degree to which harm assessment takes place depends on the premorbid history. Such a history is often absent for the children, which means that there are fewer variables that must then be extrapolated. Likewise, none of the higher-order thinking skills is fully developed. Different mechanisms govern in children, which proves a barrier to determining whether or not these functions are affected as part of the lesions.

An example of this is the differences in the functional brain networks for reading in adults and children. Reading is one of the tasks that is considered a high-level cognitive function. There are entire networks dedicated to this task that also involve various areas of the brain. Chinese characters were used as a measure of assessment. Adults had stronger connectivity within regions in the occipital lobe indicating high reliance on visual orthographic processing.

Meanwhile, children had the same high degree of inter-regional connectivity in the temporal, which showed that auditory phonological processing was dominant during their reading tasks. The adults' network systems were highly specialized to be able to tackle the different types of reading and also were distributed throughout the cortical regions. Children had more of their networks in the subcortical regions instead. It shows the evolution of reading networks that shifts from the subcortical regions to the cortical regions as it gets specialized when the child grows up and matures (Liu et al., 2018). Another source of differences can be the instances of differences when these brains are put in the same situation. A chapter in this book discussed the harmful effects of sports-related concussions. There are no separate guidelines when it comes to dealing with these concussions when it occurs in children. It persists even though there are several neuropsychological differences in the brain of children when compared with those of adults. There is a severe secondary impact of these concussions on children. There are differences in the brain tolerance of children to biomechanical forces. They are much more resistant due to a variety of reasons, such as age, different geometry of the skull, and other age-related differences. However, this also means there are more chances that the children have faced more severe forms of force (Ommaya et al., 2002). The injuries that children sustain are more likely to be exacerbated by the presence of premorbid cognitive and attention issues (McCrory et al., 2004). Any damage that is sustained during a period of rapid development is likely to cause more functional problems. That being said, there is also more significant potential for the plasticity of the affected functions when there is early brain damage. It has been confirmed by research in the case of language acquisition and development. However, there is no plasticity when it comes to cases of traumatic brain injuries (TBIs).

The argument is that there have to be sound structures present in the brain to be able to take over functions (Middleton, 2004). Plasticity is a double-edged sword in these cases. Its utility is contingent on the type of injury and the number of scar tissues involved. In cases where there is plasticity, but the nerves fail to follow the old neural pathways, it might end up causing more harm than good. It would lead to misconnections and miscommunications on a behavioral and cognitive level. Anderson et al. (2018) believed that theories of plasticity only partially explained the chances of full recovery. Some of the factors that they opined were instrumental to this explanation would be the pathology itself, the child's developmental stage and gender, and their psychosocial environment. The last factor was confirmed in the form of the impact that pathological mother-child relations might have on the mental and behavioral difficulties in preschool children. In some cases, this also meant that these kids with a low socioeconomic status had more pathological relations (Grazuleviciene et al., 2017).

Specific Developmental Disorders

Several groups of neurodevelopmental disorders can be broadly classified under abnormalities of anatomic development, genetic and chromosomal disorders, and acquired disorders. There are certain disorders that would be considered prototypical of their categories. Hydrocephalus (HC) is an example of anatomic abnormality, while Turner's syndrome (TS) is a similar example of a genetic disorder. Finally, Williams syndrome (WC) is an excellent example of an acquired disorder. A discussion about Fetal alcohol syndrome (FAS), a symptom of environmental factors: FAS is, in this aspect, completely preventable as well.

Abnormalities of Anatomic Development

Many conditions reflect the impact as a result of anatomic abnormalities. Most of these cases would have multiple causes, but the origins of most of these disorders are undetermined. It usually tends to be fatal, and even children who survive do so with a high rate of mental retardation, communication deficits, learning disabilities, and physical anomalies. HC can occur at any time during the developmental period and can disrupt both subcortical and cortical functions. The condition is a result of excessive accumulation of cerebrospinal fluid (CSF) in the brain's ventricles. This increase in the volume of CSF leads to an increase in the intracranial pressure, which puts pressure on the cerebral matter leading to their distortion. McAllister II (2012) elaborated on the pathophysiology of congenital and neonatal HC so that the causes could be understood better. Most often, both of these are caused due to neural tube defects, infection, intraventricular hemorrhage, trauma, and tumor. In either case, these result from primary genetic abnormalities and secondary injury mechanisms. Nine genes have been identified in various animal models, and so far, only one such gene has been identified in humans. Congenital forms of HC are likely to manifest in either syndromic or isolated (non-syndromic) forms. It would be easier to identify the genes responsible for the manifestation of HC when it occurs in isolated forms. The main genetic factors, in this case, would be congenital malformations, intracerebral hemorrhage, maternal alcohol use, and X-ray radiation during pregnancy. Most of the animal/rat studies conducted in this area showed that there was impaired and abnormal brain development in the early stages that was caused as a result of altered neural cells and disturbed regulation of cellular proliferation and apoptosis.

All of these factors tend to accumulate and give rise to congenital HC and its associated inflammatory reaction. However, it requires to be reiterated that these are studies that were done on animals, and there is still minimal information when it comes to the formation of human HC (Zhang et al., 2006). Most of the causes of HC are due to either blockage in the path of the CSF or issues in its resorption. One important aspect is that the age of onset determines the impact that HC will have on the brain. Despite the plasticity, as mentioned earlier, of the younger brain, the onset of HC during the in-utero phase or perinatal phase indicates a poor neurological outcome. The primary mechanism of injury during the HC is due to ventriculomegaly, which indicates the dilation of the lateral ventricles in the body. As this progresses, there are more chances of an increase in the intracranial pressure within the brain. The ventricles continue to distend, which leads to the cerebral hemispheres mold into the shape of a balloon. There are sporadic cases where there is an excess of CSF secreted within the brain. A mutation of the choroid plexuses would over-secrete the CSF. Otherwise, there is more likely to be either an obstruction or the CSF going off-path in areas such as the arachnoid spaces. The symptomatology of HC is highly individualistic and usually leads to unique manifestations based on various factors.

Dandy-Walker malformation (DWM) is a sporadic congenital disorder of the posterior fossa. The existence of a large median posterior fossa cyst that extends into the fourth ventricle, along with the upward displacement of the tentorium and lateral sinuses and a raised cerebellar vermis, would qualify for an adequate diagnosis. There is a chance that about 40% of the children with DWM would be all right intellectually. It was concluded that there are two manifestations of the DWM where there is an agenetic vermis, and the brain itself is entirely reasonable. The other type is a significant malformation that carries poor intellectual and neurological prognosis (Klein et al., 2003). It is also commonly comorbid with HC, which has also led to speculation about the real DWM. The advent of magnetic resonance imaging (MRI) has shown that there were quite a few misdiagnoses in earlier times. The pathological features of DWM are incredibly variable, and it is the degree of HC, vermian hypoplasia, and the size of the posterior fossa (Spennato et al., 2011). In most cases, using a shunt is the quickest and safest option for doing away with the excess CSF in the brain. It is a relatively simple procedure that can be done cheaply as well.

Genetic and Chromosomal Disorders

Genetic disorders can be a consequence of single-gene, chromosomal, parental imprinting, or molecular cytogenic anomalies. Genetic defects are not always hereditary in the family. They are usually transmitted through chromosomes as a consequence of their malformation, deletion, addition, or dislocation.

Turner's Syndrome

It is a disorder that predominantly affects females and is usually characterized by the absence of all or part of a healthy second sex chromosome. Its physical manifestations include congenital lymphedema, short stature, and gonadal dysgenesis. When the diagnosis occurs prenatally, there are more chances that this is done through edemas. At the same time, most girls are likely to receive this diagnosis at birth based on puffy hands and feet, which is a residual effect of cystic hygromas in utero. These are fluid-filled sacs that are found on the head or neck of a baby. These are the result of blockages in the lymphatic system. In other patients with TS, this is either diagnosed in adolescence due to failure to attain puberty or in adulthood due to repeated pregnancy. While intelligence stays unaffected, most of the patients are likely to have developmental delays that would lead to a life of special education and assistance. Deficits in the visuospatial organization, social cognition, non-verbal problem-solving, and psychomotor functioning of the patients are common. Girls with TS identify as females and are also likely to be heterosexual in orientation (Sybert & McCauley, 2004). Despite these concerns, there are many areas of uncertainty in its diagnosis and management. There are more cases of generalizations from other guidelines that deal with situations that are similar to the conditions in isolation. The future course of action would dictate that there need to be findings that are unique to TS and should be handled as such. The inclusion of patient counseling with regard to its long-term health risks and the inclusion of hormone replacement therapy would go a long way in dealing with the issues that come up with this disorder (Pinsker, 2012).

Apart from this, another disorder is holoprosencephaly (HPE). It is a brain malformation that is a result of the incomplete division of the prosencephalon into the two hemispheres. There are three levels of severity when it comes to it. The first level is where the two halves of the ventricles have been separated. The second level is where there is minimal separation, and the final stage includes a single brain ventricle that has no interhemispheric fissure. It boasts of a wide variety of anomalies where even facial midline anomalies are considered part of the HPE spectrum. It is also accompanied by developmental delays that are contingent on the severity of the brain malformation. Nearly half the patients with HPE are likely to develop epilepsy as well. Other signs include hypotonia, weakness, spasticity, and dystonia (Dubourg et al., 2007).

Acquired Disorders

There are several agents, events, and processes that can cause potential harm. Environmental toxins, radiation, infections, malnutrition, and TBI are some significant factors. These result in anomalies in the brain that usually lead to impairments and deficits.

Fetal Alcohol Syndrome

It is a condition that occurs due to prenatal exposure to alcohol. The earlier assumption was that only mothers who were binge drinkers put their babies at risk, but even social drinking can cause harm to the child. Alcohol is an established teratogen that causes variable physical and behavioral effects. Some of the mechanisms of alcohol teratogenicity are due to its degeneration of endogenous antioxidant capacity. Functional imaging of children with FAS has confirmed that there are lower levels of serotonin in the cortex and that dopamine binding has increased in the basal ganglia (Gupta et al., 2016). Some of the factors that would increase the risk of FAS development were all linked to alcohol consumption by the mothers during their pregnancy. Quantity, frequency, and timing (QFT) of alcohol usage during pregnancy. The mothers of children with FAS and fetal alcohol spectrum disorders (FASD) were more likely to drink in a binge fashion, but this would taper off three weeks before the pregnancy. Alcohol-related neurodevelopmental disability (ARND) and drinks consumed per day (DDD) (May et al., 2013). One of the main problems with the diagnosis of FAS is that it depends heavily on maternal factors. There are various confounds like body size, genetic polymorphisms, and, more importantly, paternal alcohol consumption. These are not easy to control or measure and can often skew results (Gupta et al., 2016).

Assessment and Rehabilitation

When it comes to assessing the level of functioning within children, another important aspect in addition to the factors that are usually looked at is the child's family and cultural background, general cognitive ability, and any premorbid abilities concerning their developmental age as well. All of these would require a thorough assessment in terms of clinical interviews, observation of the child, and their performance of psychometric tests as well. All of this information critically informs the diagnosis as well. The information from any of the methods in isolation would mean incorrect assumptions that would lead to significant errors. Significant sources of information when it comes to taking developmental history are pregnancy, birth events, developmental milestones, medical history, family background, and educational history. Other significant factors are also their relationships with their peers and siblings where available. These social relationships are a good insight into the development and maturity of the brain as well. In addition to these, the use of psychometric tools should be dictated concerning the cultural and familial background as well. The best information with regard would be to conduct intelligence tests like Wechsler Intelligence Scale for Children (WISC) that would be able to highlight cognitive deficits. Following this, it would be advisable to pursue a particular deficit with tools that are specifically designed for that purpose. The rehabilitation that follows would also be similar to the deficits that are already available.

Research Perspective

Neurodevelopmental disorders research is a dynamic field with a variety of areas to explore. The efforts have been made to study the neurodevelopmental impairments at different life stages. Researches have been focused on finding early age markers for the detection of disorders in young children and also the methods of intervention and therapies for the treatment of disorders. Hadders-Algra (2021) has reviewed the challenges and the opportunities in this area of research. Author has reviewed research on cerebral palsy and ADHD and has found that early diagnosis in cerebral palsy is dependent on risk factor (high or low) and on associated conditions such as intellectual disability and ADHD. As studies have shown diagnosis of ADHD in year one is very difficult as it depends more upon development of permanent brain regions, whereas for high-risk cerebral palsy cases, early detection is possible using MRI in combination with other methods. The review suggests a pivotal role of family members in early intervention in neurodevelopmental disorders which also depends upon the severity of the disorders and combination of clinical condition.

There are several neurodevelopmental disorders which have different presentation based on differences in gender and age. Study of gender-based differences is an area of research to be dealt with utmost care in population with neurodevelopmental deficit. Studies have shown sex and gender as a significant factor in deciding the psychosocial and neuropsychological differences in neurodevelopmental disorders (Bölte et al., 2023). A complex interplay of psychological, sociocultural, and biological processes is exhibited in sex-gender-based differences in such disorders. The consideration of such differences has shown a positive impact in identification and management of conditions such as ADHD in females. This indicates the need for investigation of gender differences in neural substrates of other neurodevelopmental conditions and simultaneously a need to consider these while designing the rehabilitation intervention programs. The research in this area is challenged by several factors including a multifaceted need of investigation to understand the nuances of sex-based differences.

WS is a rare multisystem neurodevelopmental disorder with specific cognitive, clinical, and social symptom presentation. Several studies have been conducted in recent decades to understand these symptoms to develop effective diagnostic and prognosis. Symptom presentation in WS is layered and may often present developmental difficulties over a period. For example, good linguistic and social abilities may hide other cognitive or clinical impairment in early stages which later becomes difficult to manage. In one of the reviews of studies on WS, authors have identified the lack of cohort studies and homogenous sample, lesser use of direct assessment methods, and investigation of similar clinical conditions such as ADHD to identify unique patterns of presentation of this syndrome. There is a need to conduct more longitudinal studies to have a clear understanding of its developmental trajectory across life span and study impairments in social cognition. These will help in overcoming the lack of intervention methods available for the patients of WS.

TS is a genetic disorder prevalent in female children born with complete or partial absence of X chromosome. Risk of cognitive and psychosocial impairments are very prominent yet understudied in females. An emerging avenue of research in TS is evaluation of efficacy of available intervention techniques to rehabilitate neuropsychological and psychosocial symptoms. There is a paucity of consistent and longitudinal studies among larger females' samples diagnosed with TS. Hutaff-Lee et al. (2019) have identified the understudied area in TS research including the interconnected presentation of various neurocognitive deficits in their review article. They have identified the need for investigation of evidence-based interventions for common deficits skills associated with TS such as executive functioning, social skills driven by executive functions, anxiety and development of valid intervention methods catering to the need of patients with TS.

Development of mobile sensing technology and its implementation in assessment of neural processing underpinning neurocognitive functioning is another avenue of research in recent times. Mobile electroencephalography (EEG) is one of the recent developments being tested as an assessment tool in neuropsychology. It may also be used for advancing and personalizing intervention techniques in neuropsychological dysfunction. A body of research has been focused on developing and testing potential uses of mobile EEG in neurodevelopmental disorders to utilize its inexpensive and non-invasive qualities. The uses of mobile EEG for children with neurodevelopmental disorders such as ADHD, cerebral palsy, learning disorders to provide flexible data acquisition, overcome cognitive and sensory sensitivity for longer durations are widely tested. In a review article authors have identified the benefits and challenges of using mobile EEG (Lau-Zhu et al., 2019). Some of the key benefits include incorporation as an assessment tool in multidisciplinary developmental studies, in longitudinal neurodevelopmental studies, ease of use by a variety of population (especially with children with clinical conditions), and assessment of neural activity in functional state. Despite several challenges, being a newly developing research tool there are challenges to be dealt with before the large-scale implementation of this method. The challenges include quality of signals, reliability of data obtained, large-scale cohort studies to validate different paradigms, development of a variety of paradigms to use in a variety of neurodevelopmental disorder treatment setups etc.

In India also neurodevelopmental research has been a key area and has seen promising progress in recent times. Studies are being conducted to investigate neurodevelopmental disorders at different age groups for development of management policies and techniques at a wider level. A large-scale study conducted by developmental researchers to identify the risk factors and prevalence of disorders and their pertaining burden showed prevalence of disorders varies at different geographical locations and found hearing impairment and intellectual disability as some of the most prominent disorders in children of the age group 2–9 years (Arora et al., 2018). They have also found children of the age group 6–9 years to be more prone to neurodevelopmental disorders. Such studies are very much needed to be done especially in developing countries to have adequate data regarding risk factors and prevalence of such disorders along with development of intervention methods that can be easily implied. Another study has identified the lack of large-scale studies, proper diagnostic and assessment tools available to all sections of society, and availability of proper intervention techniques and rehabilitation for neurodevelopmental disorders (Gaikwad & Lagala, 2021).

References

American Psychiatric Association. (2013). *Diagnostic and statistical manual of mental disorders: DSM-5* (Vol. 5, Issue 5). American Psychiatric Association, Washington, DC.

Anderson, V., Northam, E., & Wrennall, J. (2018). *Developmental neuropsychology: A clinical approach* (1st ed.). Routledge. https://doi.org/10.4324/9780203799123

Arora, N. K., Nair, M. K. C., Gulati, S., Deshmukh, V., Mohapatra, A., Mishra, D., Patel, V., Pandey, R. M., Das, B. C., Divan, G., Murthy, G. V. S., Sharma, T. D., Sapra, S., Aneja, S., Juneja, M., Reddy, S. K., Suman, P., Mukherjee, S. B., Dasgupta, R., … Vajaratkar, V. (2018). Neurodevelopmental disorders in children aged 2–9 years: Population-based burden estimates across five regions in India. *PLoS Medicine, 15*(7), e1002615. https://doi.org/10.1371/journal.pmed.1002615

Bölte, S., Neufeld, J., Marschik, P. B., Williams, Z. J., Gallagher, L., & Lai, M.-C. (2023). Sex and gender in neurodevelopmental conditions. *Nature Reviews Neurology, 19*(3), Article 3. https://doi.org/10.1038/s41582-023-00774-6

Dubourg, C., Bendavid, C., Pasquier, L., Henry, C., Odent, S., & David, V. (2007). Holoprosencephaly. *Orphanet Journal of Rare Diseases, 2*(1), 1–14.

Gaikwad, L., & Lagala, S. (2021). Prevalence and correlates of neurodevelopmental disorders among children in India: A narrative review. *International Journal of Contemporary Pediatrics, 8*(1), 200.

Grazuleviciene, R., Andrusaityte, S., Petraviciene, I., & Balseviciene, B. (2017). Impact of psychosocial environment on young children's emotional and behavioral difficulties. *International Journal of Environmental Research and Public Health, 14*(10), 1278.

Gupta, K. K., Gupta, V. K., & Shirasaka, T. (2016). An update on fetal alcohol syndrome—Pathogenesis, risks, and treatment. *Alcoholism: Clinical and Experimental Research, 40*(8), 1594–1602.

Hadders-Algra, M. (2021). Early diagnostics and early intervention in neurodevelopmental disorders—Age-dependent challenges and opportunities. *Journal of Clinical Medicine, 10*(4), Article 4. https://doi.org/10.3390/jcm10040861

Hutaff-Lee, C., Bennett, E., Howell, S., & Tartaglia, N. (2019). Clinical developmental, neuropsychological, and social–emotional features of Turner syndrome. *American Journal of Medical Genetics Part C: Seminars in Medical Genetics, 181*(1), 42–50. https://doi.org/10.1002/ajmg.c.31687

Klein, O., Pierre-Kahn, A., Boddaert, N., Parisot, D., & Brunelle, F. (2003). Dandy-Walker malformation: Prenatal diagnosis and prognosis. *Child's Nervous System, 19*(7–8), 484–489. https://doi.org/10.1007/s00381-003-0782-5

Lau-Zhu, A., Lau, M. P. H., & McLoughlin, G. (2019). Mobile EEG in research on neurodevelopmental disorders: Opportunities and challenges. *Developmental Cognitive Neuroscience, 36*, 100635. https://doi.org/10.1016/j.dcn.2019.100635

Liu, X., Gao, Y., Di, Q., Hu, J., Lu, C., Nan, Y., Booth, J. R., & Liu, L. (2018). Differences between child and adult large-scale functional brain networks for reading tasks. *Human Brain Mapping, 39*(2), 662–679.

May, P. A., Blankenship, J., Marais, A.-S., Gossage, J. P., Kalberg, W. O., Joubert, B., Cloete, M., Barnard, R., De Vries, M., & Hasken, J. (2013). Maternal alcohol consumption producing fetal alcohol spectrum disorders (FASD): Quantity, frequency, and timing of drinking. *Drug and Alcohol Dependence, 133*(2), 502–512.

McAllister II, J. P. (2012). Pathophysiology of congenital and neonatal hydrocephalus. *Seminars in Fetal and Neonatal Medicine, 17*(5), 285–294.

McCrory, P., Collie, A., Anderson, V., & Davis, G. (2004). Can we manage sport related concussion in children the same as in adults? *British Journal of Sports Medicine, 38*(5), 516–519.

Middleton, J. A. (2004). Clinical neuropsychological assessment of children. In L. H. Goldstein & J. E. McNeil (Eds.), *Clinical neuropsychology: A practical guide to assessment and management for clinicians* (Vol. 2, pp. 275–300). Wiley-Blackwell.

Ommaya, A., Goldsmith, W., & Thibault, L. (2002). Biomechanics and neuropathology of adult and paediatric head injury. *British Journal of Neurosurgery, 16*(3), 220–242.

Pinsker, J. E. (2012). Turner syndrome: Updating the paradigm of clinical care. *The Journal of Clinical Endocrinology & Metabolism, 97*(6), E994–E1003.

Spennato, P., Mirone, G., Nastro, A., Buonocore, M. C., Ruggiero, C., Trischitta, V., Aliberti, F., & Cinalli, G. (2011). Hydrocephalus in Dandy–Walker malformation. *Child's Nervous System, 27*, 1665–1681.

Sybert, V. P., & McCauley, E. (2004). Turner's syndrome. *New England Journal of Medicine, 351*(12), 1227–1238.

Weiss, B. (2000). Vulnerability of children and the developing brain to neurotoxic hazards. *Environmental Health Perspectives, 108*(suppl 3), 375–381.

Zhang, J., Williams, M. A., & Rigamonti, D. (2006). Genetics of human hydrocephalus. *Journal of Neurology, 253*, 1255–1266.

12 Neuropsychology of Epilepsy

Introduction

When it comes to epilepsy, it is essential to understand that it is one of the few disorders that brings to the fore the true nature of the brain. Despite its delicate appearance and disposition, the brain is a mixture of chemicals and electrical activity. Very few organs within the body operate at the same level as the brain. Therefore, it is not uncommon for the wires within the brain to get twisted around sometimes. It is in these moments that the delicate system within the brain gets haywire and leads to specific deficits and issues. These will be discussed in this chapter, along with the various classifications of epilepsy. The mechanisms of the disorder shall also be discussed under the section on pathophysiology. The criteria for diagnosis shall also be explored. Finally, methods of assessment, treatment, and the various myths will be deliberated as well.

Meaning of the Words and Clarity on the Terminology

When dealing with any disorder, it is crucial for clarity in terms of how the disorder is discussed. In regards to epilepsy, the International League Against Epilepsy (ILAE) decides the correct terms for the same. These were decided on the consensus surrounding the terms epileptic seizure and epilepsy. Epilepsy is "a disorder of the brain characterized by an enduring predisposition to generate epileptic seizures and by the neurobiological, cognitive, psychological, and social consequences of this condition. The definition of epilepsy requires the occurrence of at least one epileptic seizure" (Fisher et al., 2005). The epileptic seizure in question is a brief appearance of symptoms that are due to abnormal excessive neuronal activity in the brain. The term *seizure* comes from the Greek meaning to *take hold*. This term has been used in a variety of instances for any sudden event in the body and otherwise. Therefore, the term must be prefixed with epileptic for emphasis. Epilepsy is not a single condition but a diverse group of disorders. There has to be a single seizure at the very least and evidence of long-term alteration resulting from the single epileptic seizure. It makes the likelihood of the disorder epilepsy more prominent than a

DOI: 10.4324/9781032640839-15

single-event seizure (Fisher et al., 2005). The associated issues that accompany epilepsy are likely to be behavioral disturbances and the presence of problems within the interictal and postictal periods.

Moreover, there are higher chances of stigma in this regard due to its unpredictability and often public appearance in terms of motor and behavioral disturbances. Many terms are usually associated with seizures. The most common among these is the term "fits." For the sake of clarity and understanding, these terms should be disregarded; the use of the more appropriate term, epileptic seizure, would lead to better treatment rates and understanding as well.

Epidemiology

Globally, there are nearly an estimated 2.4 million people diagnosed with epilepsy. Of these, nearly 90% are found in developing regions of the world. Overall, there seems to be more risk of epilepsy in countries with lower incomes than higher incomes. However, at 1.54% and 1.03% for rural and urban prevalence, there does not seem to be a significant difference in that regard. The question of the soundness of the methodology used in these studies in epidemiological studies can take away from the reliability of the numbers being reported. It could hint at the possibility of diseases secondary to epilepsy. Developed countries seem to follow a U-shaped curve, where the children and the elderly are more likely to be affected. However, the opposite seems to be true in developing countries, where the disorder seems to peak in early adulthood (A. Singh & Trevick, 2016). Exploring one of the most prominent developing countries in the world: India, might give a fair picture of the ground reality in this situation. Many studies have been conducted about epilepsy and its prevalence in India, which have also seemed summarised in metanalyses and reviews. Figures estimate that approximately 10 million people in the country might have some form of epilepsy. The age of onset is that most cases would be around 15. However, there was a rise in the trend where the symptoms are more likely to manifest in the second or third decade of the individual. Males were more likely to be affected than females, while the place of residence (urban and rural) did not seem to be affected.

In most cases, the epileptic seizure was attributed to supernatural causes. It directly played into their cultural beliefs as well. The incidence of mental retardation was at 1.25%, and only in 7% of cases; there was behavioral difficulty observed. One worrying scenario was that there was a huge treatment gap. Many of the people with epilepsy did not seek treatment, and this was primarily due to a lack of knowledge, a level of acceptance concerning its cultural origins, and finally, the absence of trained mental health professionals. A phenomenon known as Hot Water Epilepsy was observed in the southern parts of the country. It was necessarily brought on by pouring hot water on the head, which led to epileptic seizures; whenever there is a group of people that are not getting the appropriate treatment for a condition, that is known as a treatment gap. There are two components when it comes to the treatment gap. Some people have access, and some do not. Even among those who do have access to treatment, their adherence to the treatment that is prescribed also contributes to the gap in treatment for that particular disorder. In part, due to the cultural aspects surrounding epilepsy, the first person that is usually contacted is the faith healer rather than a doctor. The stigma surrounding the condition stems directly from the fact that there is no knowledge of its etiology. In cases where this is understood, there is more likely to be discrimination against the person who is suffering from it. The fact that epilepsy like a physical disorder can be treated through medication is something that is still lost on many people (Pal et al., 2010; Panagariya et al., 2019; Satishchandra et al., 2014).

Classification

Many types of epileptic seizures depend on where the seizure occurs, along with its etiology. This classification recently underwent a review by the ILAE. It is an operational revision and has also clarified the nomenclature to be used while defining the same. The classification of seizures has dated till the time of Hippocrates, the term itself having originated in Ancient Greece. There has been various framework that has defined the terms used as well. The previous attempts were based on the anatomy with the temporal, frontal, parietal, occipital, diencephalon, or brainstem seizures. The current classification, as described below, is not hierarchical. The first course of action in classifying seizures is based on whether the seizure is focal or generalized. The onset of the seizure could also be linked to its manifestation. The terms focal and generalized also denote the onset of the same. Focal seizures are characterized by the level of awareness retained of the surroundings. When the subject retains awareness of self and environment, it would mean that the aware focal seizure without other classifiers would be a simple partial seizure. The absence of such awareness would correspond to a complex partial seizure. Moreover, these seizures would also be classified based on motor or non-motor signs.

Whether the epileptic seizure is generalized or focal is based on the anatomy, while the motor and nonmotor signs would mean that there is a behavioral aspect to it. The level of awareness, while being a suitable method of classification between simple and complex seizures, might not always be freely available. Moreover, even the presence of motor symptoms tends to be slight and easy to miss. Additional classifiers like sensory/motor, tonic/clonic, and automatism would only serve to expand our understanding of the seizure. Terms that were discontinued were the simple/complex seizures that were used as classifiers for partial seizures. Also, these terms are loaded with judgments that might be misunderstood by the general public. Another term that has been discontinued is convulsions. The term is popular but also ambiguous. While the term is often considered synonymous with seizure, it has been excluded from the 2017 classification. Despite this, it might still be a popular term. One of the main differences from the earlier classification is that there are various types of focal seizures like automatisms, behavior arrest, cognitive, emotional, and sensory, along with the tonic-clonic and myoclonic classifications. There has been an expansion of seizure types under both focal and generalized seizures (Fisher et al., 2017).

Pathophysiology

Epileptic seizures arise from excessive synchronous and continuous release of a group of neurons. The main feature of these syndromes is that there is a continued increase in the strength of this exciting neuronal activity. Some of the primary forms of causes would include anoxia, trauma, tumors, and infection, among others. There are also cases where there are no actual causes determined (Engelborghs et al., 2000). In the following sections, genetics, kindling, epileptogenesis, ictogenesis, and neurochemical transmissions will be discussed

Genetic

Nearly half of the patients who have epilepsy have a familial background in it. There are cases where familial epilepsies like juvenile myoclonic epilepsy, benign childhood epilepsy, and others lead to an interaction between the genes and environmental factors. Thomas and Berkovic (2014) argue against the predominant view that while there has been much research that has identified aspects of genetics and its influence on epilepsy, these have been relatively wasteful in the field and cases of diagnoses. The hidden genetics forms the basis of the disorder and would help propel treatment and management of the disorder. The genetics do not work in isolation and would require activation based on the external factors for its expression. The molecular background of these disorders has ensured that they are easier to treat as well. A significant portion of generalized genetic epilepsies and familial temporal lobes do not follow Mendelian patterns of inheritance. Even epilepsies that were acquired through trauma, stroke, infections, or congenital malformations were found to harbor genetic contributions within them. Genetic studies have contributed to the loss of the term idiopathic seizures. With the advent of the MRI, lesions that were too small to be detected by CT scans were determined, which led to several unexplainable seizures as having a determined basis. There is not a lot of presence of the disorder in first-level relatives as epilepsy is not a Mendelian disorder. It would mean that there are *de novo* mutations for epilepsies. These mutations are genetic alterations that are present for the first time in a family member.

Rather than a monogenic basis, epilepsy tends to be polygenic and complicated. De novo mutations are more likely to be implicated in the genetic mechanisms of epilepsy as it has also been found in psychiatric disorders like schizophrenia, autism spectrum disorders, and intellectual deficits. Massively parallel sequencing (MPS) technologies have been at the forefront of introducing and identifying genes that might be responsible for the expression of epilepsy. This software has been optimized for DNA diagnostics. These are popular over traditional sequences as it has a high level of screening and is sensitive enough to detect all varieties of mutations as well. Finally, it is reliable and cost-effective too (Hildebrand et al., 2013). Genes are the future in terms of most disorders, and therefore, their intricate understanding would determine the future of diagnosis and the methods for its prevention.

The earliest model that seemed to explain a mechanism of epilepsy was called the kindling effect or the kindling model. It was done first in rats where continuous electrical stimulation did not seem to have a perceptible effect, however, after a few trials there seemed to be convulsions in the forelimbs, despite the intensity being constant through the number of trials. It was dubbed as the kindling effect where there is an alteration in the brain that occurs through continuous stimulation. It is not the first time that there has been an electrical stimulation that was applied to the brain. The first was done by Fritsch and Hitzig, where they applied electrical stimulation to a frog's severed leg that showed that electricity indeed seemed to be the medium that was used within the body to convey messages. There is a considerable similarity between kindling and epilepsy. In particular, it would seem like kindling is experimental epilepsy. Hence, it would form a well-controlled environment to test various chemicals and factors for their susceptibility to convulsions (Gaito, 1974).

Neurochemical

Neurochemicals are often the key to communication and mechanisms of natural change within the brain and body. Therefore, any dysfunction would imply the advent of disorders and diseases as well.

Gamma-aminobutyric acid (GABA) has been well-established as the primary inhibitory neurotransmitter. GABA synapses are present in nearly all brain structures, especially in the cerebral cortex. Given that epileptic seizures are a result of overexcited neurons, it would follow that the presence or absence of GABA would indicate the effects of epileptic seizures as well. GABAergic inhibition is defined as "a set of variables distributed in multiple inhibitory pathways. Because the values of these variables are continuously changing and interacting, the state of inhibition can only be approached when the system is frozen in a quasi-stable state" (Bernard et al., 2000). Its effects on the factors are somewhat differentiated and dependent on time, area, and pathway. Some of the changes that typically follow a GABAergic reaction are epileptic, while others are anti-epileptic. If there are limited GABA-ergic reactions, the chances of epilepsy will increase, and if its effects are enhanced, then there will be a decrease in the likelihood of epilepsy. Before surgery, many preoperative and intraoperative tests are carried out to determine the epileptogenic areas. The primary use of GABA comes from the inhibitory postsynaptic potentials (IPSPs). Whenever there is altered GABA-mediated inhibition in the human cortex, there are chances that epileptic seizures are likely to occur. There could be bursts of neural activity with even a small decrease in inhibitory action (Tasker & Dudek, 1991).

Glutamate and glutamatergic synapses play a critical role in most cases as far as epilepsy is concerned. Chapman (2000) believes that there are seizures that are provoked in animals due to a wide range of glutamatergic molecular mechanisms. Glutamate is a potent excitatory neurotransmitter in the brain. One of the mechanisms through which glutamate tends to start its excitatory action is by the ligand-gated ion that increases sodium and calcium conductance. Intracellular recordings during epilepsy have shown that there is a paroxysmal depolarising shift that leads to membrane spikes. It is similar to a substantial excitatory synaptic potential. Whenever there are chances of alterations within the glutamate receptors, the chances of epileptogenesis increase accordingly. Release of glutamate has been observed in cases where there was epilepsy evoked and also during temporal lobe epilepsy. Barker-Haliski and White (2015) reviewed the evidence that directly connected glutamatergic mechanisms with seizures and epilepsy. There could be a case where it is difficult to determine whether it is the abnormal signaling that is caused by seizures or the other way around. There have been cases where the essential excitatory neurotransmitter has been implicated in its role in epileptic seizures. An understanding of these mechanisms would lead to the prevention of epileptogenesis through control. Glial cells that are traditionally supposed to be the support neurons are implicated in controlling the glutamate mechanism. Astrocytes tend to function during synaptic signaling by way of neurotransmitter transporters and receptors. Astrocytes, in particular, contain the GLAST and GLT-1 that are necessary for the reuptake of glutamate at the synapses. Even microglial cells play a similar role in this regard. One particular gene defect at the alpha4 subunit of the nicotinic cholinergic receptor could be influencing glutamate release. The altered expression or function of glutamate receptors may contribute to the occurrence of epilepsy. There are similar molecular changes in the human hippocampus removed from patients with focal seizures and kindled rats (Meldrum et al., 1999).

Infection

The brain is situated in a very controlled environment where there are not many chances for foreign particles to enter. There is research evidence that shows that people with a history of central nervous system infection are at a higher risk of developing epilepsy than people who do not. The age at which the infection is acquired has also been linked with the type of epilepsy as well as the time between the occurrence of epilepsy and the infection. Also, the presence of a history of infection does not negatively impact outcomes for epilepsy surgery. Hence, these should not be considered when patients seek surgery for the same (Lancman & Morris III, 1996). One of the significant disorders that tend to affect children is known as febrile infection-related epilepsy syndrome (FIRES). A previously healthy child who might have had a short febrile illness would also lead to an onset of status epilepticus (SE). These conditions usually lead to a diagnosis of FIRES. There are low levels of the disorder in the global scenario. It has been attributed to its misdiagnosis and underdiagnosis as well, in which case the real figures might be much higher. When it comes to diagnosis, there are no known tests that are available specifically for FIRES. Some of the significant methods for identifying FIRES are checking the CSF for infectious agents. MRI and EEG are also conclusive systems that might lead to an understanding of the progress and patterns of the disorder (van Baalen et al., 2017).

FIRES is an epileptic syndrome that tends to affect children who are aged 3–15 years. A case study would be able to highlight some of the challenges and options that are usually present in similar cases. A status epilepticus is a condition that results from the failure of systems that are responsible for the termination of abnormally long seizures. Most cases occur when there is an unremarkable febrile illness. The six-year-old female presented with a case of a tonic-clonic seizure. There was no case of epilepsy or any other health problems in history. She had reported that she had not been feeling well a week before admission but did not have a fever at present. The patient had been unresponsive on arrival. The MRI and CT scans were clear. EEG showed that there were generalized periodic epileptiform discharges. EEG was also used to monitor the patient when there was increased seizure activity. The patient was treated for possible infections and autoimmune diseases. The patient went on to have moderately severe global encephalopathy with ID and intractable epilepsy (Fox et al., 2017).

There is a general level of acceptability regarding the role that neurotransmitter systems and ion channels play in neuronal excitability, which in turn contributes to epileptogenesis. This increased knowledge on the subject would lead to a better understanding of the phenomenon as well as lead to effective methods of treatment and management (Engelborghs et al., 2000).

Diagnosis

Patients with epilepsy are easily diagnosable with the help of structural neuro-imaging, especially in cases where there is an onset of focal seizures above the age of 20. Most patients who exhibit signs of epilepsy are likely to have undergone a CT scan; however, it is only useful when there are mass lesions present. The MRI can be used for lesions that are small and may often escape a regular CT scan in this regard. These are likely to be lesions that are non-progressive and easily treated with surgery should medications fail (Mellers, 2004). When it comes to epilepsy, the chances that epilepsy is misdiagnosed tend to be pretty standard. Epilepsy is a syndrome, and there are more chances that cases of psychogenic epilepsy are often falsely diagnosed as epilepsy itself. The over-interpretation of EEG usually leads to misdiagnosis in this regard. Video EEG is considered to be the gold standard when it comes to a diagnosis of epilepsy. This method is especially considered when there are continued attacks that occur despite medication. The principle of the vEEG is to record episodes that can demonstrate that there are no changes that occur in EEG during the clinical event. The other condition that is usually often misdiagnosed as epilepsy is syncope. It is mostly due to the common misconception that seizures usually lead to loss of consciousness. Other disorders that are usually diagnosed as epilepsy are hypoglycemia, panic attacks, and paroxysmal movement disorders (Benbadis, 2009). All of this points to a lack of understanding of epilepsy and how these are manifested. Van Donselaar et al. (2006) and the Dutch Study Group of Epilepsy in Childhood have called into question the confidence of these diagnoses. This belief stems from the fact that most cases of first seizures tend to be based on eyewitness accounts of the closest relatives at the time. While it might seem that interpreting EEG data would be a straightforward affair, there would be differences in terms of other observers as well. Also, not most cases may be admitted and undergo video EEG so that there is better clarity in terms of the disorder. Whenever there is a case where a patient comes in with a report of a seizure, the very first job is to figure out if it is epileptic. As discussed previously, there is a high rate of misdiagnosis if the decision is made solely based on the tonic-clonic movement. Syncope is a disorder that also has a similar behavioral manifestation to epileptic seizures. Even if an EEG recording is done, there are cases where there is an absence of epileptiform. Even when there are epileptiform abnormalities in the EEG report, there is a slight chance of this developing into epilepsy. EEG also might give results that might contradict clinical observation. There might be a clinical diagnosis of focal seizure, while the EEG might give the epileptiform of generalized seizure. There is a general overreliance on clinical evidence like the EEG, and the diagnosis might be made purely on the EEG itself. Diagnosis of epilepsy based on a single seizure is subjective and might need interobserver reliability. The confidence of a proper diagnosis would derive from good eyewitness accounts. There are many cases, and with the advent of mobile technology that, video proof is also available. The general population is tech-savvy and understands that it might be able to speed up the process by giving video proof. In order to avoid misdiagnosis in these cases, the possibility of long-term video EEG should also be entertained. The anti-epileptic drugs (AEDs) that are typically prescribed are, in some cases, harmful and are likely to give epilepsy if prescribed in cases where there are none. This effect will be elaborated in the section that deals with the AEDs. It is a word of caution to specialists who might prescribe AEDs as a method of ascertaining the diagnosis. The possible socio-economic ramifications, along with the side effects of the medications, mean that the prescription should be avoided until a proper diagnosis is met. Zanzmera et al. (2019) have brought up the possibility of replacing long-term EEG with short-term EEG. This possibility was particularly to avoid the misdiagnosis of psychogenic nonepileptic seizures (PNES). Long-term video EEG (LTVEEG) has become the gold standard to avoid misdiagnosis, but in countries like India where there is a paucity of resources, there is a need to explore whether short-term video EEG (STVEEG) is just as useful. LTVEEG typically lasts from a few hours to several days. It also includes recording during sleep and 24 hours of monitoring.

In contrast, the authors decided that STVEEG would last anywhere between 40 and 120 minutes. Nearly 80% of the sample gave an adequate diagnosis of PNES, while the remaining had to undergo the LTVEEG for the same. This result shows that STVEEG could be used in conditions where there are strained resources.

The main crux of this section will now be summed up using the latest evidence-based diagnostic guidelines for the management of epilepsy. Evaluation should typically start with an EEG in cases where there is an unprovoked seizure, as it is going to give the likelihood of a recurrence. Brain imaging like CT scans and MRIs should also be ordered. Clinicians are also advised to take an AED immediately to reduce the attack of a second seizure (S. P. Singh et al., 2017).

Neuropsychological Assessment in Epilepsy

1	Motor Functions	
	Motor Speed	a. Finger Tapping Test
	Motor Coordination	a. Hand Tapping
	Mental Speed	a. Digit Symbol Substitution Test
2	Attention	
	Sustained Attention	a. Colour Cancellation
		b. Digit Vigilance Test
	Focused Attention	a. Color Trails Tests
		b. Clinical Observation
	Divided Attention	a. Triads Test
3	Expressive Speech	
	Repetitive Speech	a. Repeating Sounds
		b. Repeating Words
	Nominative Speech	a. Categorical Naming
		b. Object Naming
	Narrative Speech	a. Sentence Construction
4	Executive Functions	
	Verbal Fluency	a. Phonemic Fluency
	Design Fluency	a. Design Fluency
		b. N-Back Verbal
	Verbal Working Memory	a. VSWM Span Task
		b. Self-Ordered Pointing Test
	Visuospatial Working Memory	a. N-Back Visual
		b. Self-Ordered Pointing Test
	Planning	a. Porteus Maze
		b. Tower of London
	Set Shifting	a. Wisconsin Card Sorting Test
	Response Inhibition	a. Stroop Test
	Motivation	a. Clinical Observation
	Behaviour Change	a. Parental Interview

5 Visuo-Perceptual Ability	
General Ability	a. Motor-Free Visual Perception Test
	b. Bender-Gestalt Test
Visuo-Conceptual Ability	a. Picture Completion
Visuo-Constructive Ability	a. Block Design
	b. Complex Figure Test
Visual Recognition	a. Recognition
Apraxia	a. Symbolic and Sequential Acts
6 Somatosensory Perception	
Tactile Finger Localisation	a. Finger Localisation
Reading	a. Reading Comprehension
Writing	a. Writing to dictation
	b. Copying
7 Mental Ability	
Calculation	a. Appropriate Age Sums
Verbal Comprehension	a. Token Test
Verbal Learning and Memory	a. Rey's Auditory Verbal Learning Test (RVLT)
Visual Learning and Memory	a. Memory for designs
	b. Complex Figure Test

All of these procedures are important when it comes to assessing the deficits and also regarding whether the patients are good candidates for surgery. Sayuthi et al. (2009) provided evidence that showed that neuropsychological testing in epilepsy surgery patients demonstrates how the tests were broadly used to assess the effect of the surgery on the patient's cognitive function and quality of life. Some of the tests that were used were the Wechsler Adult Intelligence Scale-III (WAIS-III), the Wechsler Memory Scale, and the QOLIE 31. Epilepsy surgery is currently cleared only for the frontal lobe and the temporal lobe, which are primarily responsible for executive functioning and memory, respectively. These tests in the preoperative stage were useful in determining the laterality of epilepsy and its focus. The neuropsychological test is vital in terms of investigations that provide evidence beyond the machine findings. The areas of the brain that need to be assessed also depend on the type of epilepsy that has been confirmed by the EEG and CT scans. Cognitive impairment can be caused due to epilepsy but there are also cases where it might precede the same as well. Neuropsychology and neuropsychologists have moved on from their role of being the first source of diagnosis in the absence of adequate brain imaging techniques. Surgical target regions that are assessed by using many modalities of information and especially sensitive neuropsychological tests have reduced the resection of non-affected functional brain areas (Helmstaedter & Witt, 2017).

Neurocognitive and Behavioural

Deficits in cognitive functions are observed in patients with epilepsy with alarming frequency. The cause of epilepsy, along with its severity and the brain site, are the significant factors that decide the type of cognitive deficits. The nature of the therapy being used along with the number of anti-convulsants is also key to understanding the type of cognitive deficits. Most epilepsy-related deficits are better assessed within the developmental stages of the patients. In epilepsy, there might be a decline in the IQ, which is often accompanied by decline and deficits in attention, psychomotor speed, learning, memory, visuospatial skills, and executive functioning. There are more cases of generalized epilepsy leading to a more severe level of deficits than focal seizures. The list provided above is comprehensive enough to be able to identify most of the deficits. These are all often part of a neuropsychological battery. Administration usually can take up to 4–6 hours depending on the speed of the subject. However, it will provide an exhaustive profile of the patient's neuropsychological skills (Verma et al., 2011).

Treatment

There are 70% of adults with new-onset epilepsy who are likely to be receptive to the AEDs. However, half of them will also suffer from acute side effects. While there are newer AEDs that have brought in more treatment options and ease of usage and access, there is a need to revitalize the search in order to find a solution to drug-resistant epilepsy. Anti-epileptogenic compounds that prevent epilepsy before the occurrence of the first seizure will go a long way in addressing the treatment gap in epilepsy (Schmidt & Schachter, 2014).

In most cases for treatment, once there has been a diagnosis of epilepsy are AEDs. These are instrumental in adjusting the mechanisms that were previously discussed in the chapter. However, there are many side effects to the drugs, and also, there is a growing minority of people who are resistant to the treatment itself. There are several new areas and strategies are proposed as proof-of-principle. Some of these include gene transfer and even cell-transplantation approaches. Some of the viral vectors that were under consideration were the adeno-associated virus and neuropeptide by gene therapy. These are also comparable to stem-cell research (Sørensen & Kokaia, 2013). Treatment options have come a long way from just the first use of bromides.

There are now multiple treatment modalities that include implantable devices as well as surgical options. However, there are increased chances of suicidal ideation while on these medications, and there should be a requirement for screening for the same. Epilepsy surgery is usually advised to post the medications. This option is usually suggested in extreme cases where there are patients with complicated cases of focal seizures or generalized seizures. Some surgeries are often done, but there is no concrete evidence at the time. Therefore there is a requirement that the patients are adequately counseled on the risks and benefits of opting for the surgery as opposed to continuing with the pharmacotherapy. Another type of epilepsy surgery is corpus callosotomy and hemispherectomy. Finally, vagal nerve stimulation is also a proper treatment plan for patients that are not responsive to medication and also are not good candidates for surgery. It is approved for patients above the age of 12. However, there are no optimal settings for the VNS procedure.

Therefore, the therapy is always started at the lowest settings and then adjusted depending on the effects during each visit S. P. Singh et al. (2017). When the patient meets the criteria for surgery, which includes factors such as non-response to AED, accessibility of the area, absence of confounding medical conditions, and improvement of the Quality of Living, then the patient is counseled on the effects of the surgery. For this purpose, an epilepsy surgery counselor who might also be a neuropsychologist is referred. This option would enable the patient and their family to discuss any concerns that they might have about the surgery and can make an informed decision regarding the same. The counselor should try to find out as much as possible about the patient's social support and family systems. Some of the significant misunderstandings and concerns might be related to the procedures before the surgery, the duration of stay after the surgery, and how the patient might feel during the surgery. The counselor is also responsible for giving post-operative care and advice as well. During the same period, it must be made clear that there are chances of the patient suffering from emotional problems after the surgery. There might be tiredness, loss of sleep, and reduced appetite for about a month and a half post the surgery. All of these symptoms are likely to resolve on their own with no intervention. However, some patients might feel the need to take medications to deal with them as well – the role of the family members in reporting any unusual behavior as soon as possible. Finally, the medications are necessary up to at least a year after the surgery. The benefits of the surgery itself would only be evident about two years after.

Myths of Epilepsy

Several myths surround the disorder of epilepsy, and we will try to disprove some of the more common ones.

1 **You can swallow your tongue during a seizure.**
   This event is not physically possible. The tongue is attached and cannot be swallowed under any circumstances.
2 **You should force something into the mouth of someone having a seizure.**
   This would only lead to them damaging their teeth or gums. The only thing to be done is that a soft pillow should be placed under their head to protect it against the floor.
3 **You should restrain someone from having a seizure.**
   This will be counter-productive to your intentions. Trying to restrain someone during epilepsy would only lead to more injuries. The epileptic seizure should just run its course.
4 **Epilepsy is contagious.**

Epilepsy is a condition that is limited to the brain and excessive activities, and it is not contagious through any medium.

5 **Only kids get epilepsy.**
Not true. There is no restriction when it comes to epilepsy; even the geri-atric population is likely to get it. The main difference being that the seizures in the elderly are often due to other problems like stroke or heart disease.

6 **People with epilepsy are disabled and cannot work.**
It is not a disability. Many people can carry on with their lives without having seizures. The severity of the attacks is a significant determiner of their employability.

7 **With today's medication, epilepsy is pretty much solved.**
While medical research has come a long way, there are still many issues that cannot be solved simply by popping a pill. Several strains of epilepsy might be resistant to medications. There are also several people whose medication does not work.

8 **You cannot die from epilepsy.**
Prolonged cases of status epilepticus have been the cause of death in some cases.

9 **You cannot tell what a person might do during epilepsy.**
While the behavior exhibited by a person undergoing an epileptic attack is inappropriate for the time. This behavior is not controlled and cannot cause harm intentionally.

10 **People with epilepsy are physically limited in what they can do.**
This is not true in all cases. There are many people with epilepsy who are still able to carry on with their lives despite the illness.

Research Perspective

Epilepsy is one of the most common multifaceted neurologic conditions. Along with its detailed symptom presentation, it can have psychiatric comorbidity, which requires a combination of management techniques which is often very complicated and burdensome. Neuropsychological research in identifying a better understanding of the disease and its comorbidities, treatment gaps, de-veloping better intervention plans, and epileptic surgery has grown in recent decades. Researchers have been focusing on neurodegenerative proteins and their role in the progression of epilepsy, the assessment of factors like language and social functions, which have an underlying association with epilepsy pres-entation, the association of epilepsy with factors such as exercising, cogni-tive functions, vascular abnormalities, and metabolic abnormalities (Hermann et al., 2017). Researchers have been using tools such as genetics, neuroimag-ing, digital tools, and non-invasive brain stimulation techniques for reaching a broader understanding of the disease. Differences in epilepsy presentation and epileptic care based on gender, culture, cast, socio-economic status, etc., shall also be studied on larger samples in various settings as we progress with technical advances. They play a significant role in how epilepsy and its comorbidities are treated.

Another area of research that has received scant attention is suicidal tendencies in epilepsy patients. A comprehensive set of literature indicates that suicide rates are higher in epilepsy patients than in the average population, but their neuropsychological profile is still under studied (Harcourt, 2020). Association between suicidal thoughts and epilepsy features, including frequency, type, location, presentation, and category, are some of the vital areas to be considered for further research in the future. The unclear picture of these relationships have made it harder for the assessment and management of suicidal tendency which needs to be studied further.

Genetic research in epilepsy is another trending area of research. The use of genetics can prove to be a feasible and cost-effective method for diagnosis, for understanding the root causes of the disease, and hence, in the effective development of treatment and intervention plans (Myers et al., 2019).

Digital tools for the management of epilepsy are an emerging area of research. Studies have been conducted to develop practical digital tools to overcome common hindrances in epilepsy management, including lack of information, availability or access to experts, and issues in self-management of the disease. Often epilepsy patients have clinical morbidity, which makes treatment and intervention difficult. Shegog et al. (2020) have proposed a digital ecosystem for an efficient implementation of digital technology. Ecosystem tools manage three dimensions: self-management education tools, monitoring devices for seizure detection, and day-to-day monitoring (symptoms, etc.) devices. They propose that a digital ecosystem can help design effective interventions for each patient's overall health. The prospect of digital tools implementation is high, which needs to be substantiated with awareness programs for the effectiveness of the tools, research to prove the accuracy of data, the effectiveness of tools, financial feasibility, and other related issues.

In a review of social cognitive abilities, including emotion regulation and theory of mind in epileptic patients, researchers studied social cognitive abilities and identified the gaps in assessment methods used in clinical setups (Eicher & Jokeit, 2022). They recommend the development of social-cognitive ability assessment tools with ecological validity to be used in research and clinical practices. They have also identified the need to have longitudinal studies on larger samples to study the effect of different features of epilepsy, like frequency, type, duration, and onset, on subsets of social cognition like emotion regulation, cognitive abilities, and executive functions. This can help in widening our understanding of the role of social-cognitive abilities in epilepsy and the development of effective intervention techniques.

The self-care and management of caregivers in diseases like epilepsy are of equal importance. Since epilepsy is still misinformed in Asian countries and due to a lack of support, and access to technical support, family members end up taking care of the patients. Also, taking care with limited information and resources often tends to be very exhausting. A review of caregiver studies in India was conducted, and it was found that their mental and physical health, financial condition, and family conditions are adversely affected by caregiver burden (Bapat & Shankar, 2021). This review has identified some of the major limitations and scope for further exploration in caregiver research. The factors to be included in future studies are mental, physical health, and psychosocial health of caregivers, cross-cultural differences, distinction in rural and urban settings, etc.

References

Bapat, D. A., & Shankar, A. (2021). A review of caregiver distress in epilepsy in India: Current issues and future directions for research. *Epilepsy & Behavior, 116*, 107787. https://doi.org/10.1016/j.yebeh.2021.107787

Barker-Haliski, M., & White, H. S. (2015). Glutamatergic mechanisms associated with seizures and epilepsy. *Cold Spring Harbor Perspectives in Medicine, 5*(8), a022863.

Benbadis, S. (2009). The differential diagnosis of epilepsy: A critical review. *Epilepsy & Behavior, 15*(1), 15–21.

Bernard, C., Cossart, R., Hirsch, J., Esclapez, M., & Ben-Ari, Y. (2000). What is GABAergic inhibition? How is it modified in epilepsy? *Epilepsia, 41*, S90–S95.

Chapman, A. G. (2000). Glutamate and epilepsy. *The Journal of Nutrition, 130*(4), 1043S–1045S.

Eicher, M., & Jokeit, H. (2022). Toward social neuropsychology of epilepsy: A meta-analysis on social cognition in epilepsy phenotypes and a critical narrative review on assessment methods. *Acta Epileptologica, 4*(1), 24. https://doi.org/10.1186/s42494-022-00093-1

Engelborghs, S., D'Hooge, R., & De Deyn, P. (2000). Pathophysiology of epilepsy. *Acta Neurologica Belgica, 100*(4), 201–213.

Fisher, R. S., Boas, W. V. E., Blume, W., Elger, C., Genton, P., Lee, P., & Engel Jr, J. (2005). Epileptic seizures and epilepsy: Definitions proposed by the International League Against Epilepsy (ILAE) and the International Bureau for Epilepsy (IBE). *Epilepsia, 46*(4), 470–472.

Fisher, R. S., Cross, J. H., French, J. A., Higurashi, N., Hirsch, E., Jansen, F. E., Lagae, L., Moshé, S. L., Peltola, J., & Roulet Perez, E. (2017). Operational classification of seizure types by the International League Against Epilepsy: Position Paper of the ILAE Commission for Classification and Terminology. *Epilepsia, 58*(4), 522–530.

Fox, K., Wells, M. E., Tennison, M., & Vaughn, B. (2017). Febrile infection-related epilepsy syndrome (FIRES): A literature review and case study. *The Neurodiagnostic Journal, 57*(3), 224–233.

Gaito, J. (1974). The kindling effect. *Physiological Psychology, 2*(1), 45–50.

Harcourt, S. (2020). The neuropsychology of epilepsy and suicide: A review. *Aggression and Violent Behavior, 54*, 101411. https://doi.org/10.1016/j.avb.2020.101411

Helmstaedter, C., & Witt, J.-A. (2017). How neuropsychology can improve the care of individual patients with epilepsy. Looking back and into the future. *Seizure, 44*, 113–120.

Hermann, B., Loring, D. W., & Wilson, S. (2017). Paradigm shifts in the neuropsychology of epilepsy. *Journal of the International Neuropsychological Society, 23*(9–10), 791–805. https://doi.org/10.1017/S1355617717000650

Hildebrand, M. S., Dahl, H.-H. M., Damiano, J. A., Smith, R. J., Scheffer, I. E., & Berkovic, S. F. (2013). Recent advances in the molecular genetics of epilepsy. *Journal of Medical Genetics, 50*(5), 271–279.

Lancman, M. E., & Morris III, H. H. (1996). Epilepsy after central nervous system infection: Clinical characteristics and outcome after epilepsy surgery. *Epilepsy Research*, *25*(3), 285–290.

Meldrum, B., Akbar, M., & Chapman, A. (1999). Glutamate receptors and transporters in genetic and acquired models of epilepsy. *Epilepsy Research*, *36*(2–3), 189–204.

Mellers, J. D. C. (2004). Neurological investigations. In L. H. Goldstein & J. E. McNeil (Eds.), *Clinical neuropsychology: A practical guide to assessment and management for clinicians* (Vol. 2, pp. 57–78). Wiley-Blackwell.

Myers, K. A., Johnstone, D. l., & Dyment, D. a. (2019). Epilepsy genetics: Current knowledge, applications, and future directions. *Clinical Genetics*, *95*(1), 95–111. https://doi.org/10.1111/cge.13414

Pal, S. K., Sharma, K., Prabhakar, S., & Pathak, A. (2010). Neuroepidemiology of epilepsy in Northwest India. *Annals of Neurosciences*, *17*(4), 160.

Panagariya, A., Sharma, B., Dubey, P., Satija, V., & Rathore, M. (2019). Prevalence, demographic profile, and psychological aspects of epilepsy in North-Western India: A community-based observational study. *Annals of Neurosciences*, *25*(4), 177–186.

Satishchandra, P., Santhosh, N., & Sinha, S. (2014). Epilepsy: Indian perspective. *Ann Indian Acad Neurol*, *17*(5), 3.

Sayuthi, S., Tharakan, J., Pieter, M. S., Mar, W., Madhavan, M., Tahir, A., & George, J. (2009). Neuropsychological assessment in epilepsy surgery–preliminary experience in a rural tertiary care hospital in north east Malaysia. *The Malaysian Journal of Medical Sciences: MJMS*, *16*(1), 39.

Schmidt, D., & Schachter, S. C. (2014). Drug treatment of epilepsy in adults. *BMJ*, *348*, 254.

Shegog, R., Braverman, L., & Hixson, J. D. (2020). Digital and technological opportunities in epilepsy: Toward a digital ecosystem for enhanced epilepsy management. *Epilepsy & Behavior*, *102*, 106663. https://doi.org/10.1016/j.yebeh.2019.106663

Singh, A., & Trevick, S. (2016). The epidemiology of global epilepsy. *Neurologic Clinics*, *34*(4), 837–847.

Singh, S. P., Sankaraneni, R., & Antony, A. R. (2017). Evidence-based guidelines for the management of epilepsy. *Neurology India*, *65*(7), 6.

Sørensen, A. T., & Kokaia, M. (2013). Novel approaches to epilepsy treatment. *Epilepsia*, *54*(1), 1–10.

Tasker, J. G., & Dudek, F. E. (1991). Electrophysiology of GABA-mediated synaptic transmission and possible roles in epilepsy. *Neurochemical Research*, *16*, 251–262.

Thomas, R. H., & Berkovic, S. F. (2014). The hidden genetics of epilepsy—A clinically important new paradigm. *Nature Reviews Neurology*, *10*(5), 283–292.

van Baalen, A., Vezzani, A., Häusler, M., & Kluger, G. (2017). Febrile infection–related epilepsy syndrome: Clinical review and hypotheses of epileptogenesis. *Neuropediatrics*, *48*(01), 005–018.

Van Donselaar, C. A., Stroink, H., Arts, W., & Dutch Study Group of Epilepsy in Childhood. (2006). How confident are we of the diagnosis of epilepsy? *Epilepsia*, *47*, 9–13.

Verma, N., Singh, N., Malhotra, S., & Sharma, V. (2011). Neuropsychological assessment and management of patients with epilepsy: A practical approach. In V. Sharma & S. Malhotra (Eds.), *Clinical neuropsychology* (pp. 151–171). Harprasad Institute of Behavioral Sciences.

Zanzmera, P., Sharma, A., Bhatt, K., Patel, T., Luhar, M., Modi, A., & Jani, V. (2019). Can short-term video-EEG substitute long-term video-EEG monitoring in psychogenic nonepileptic seizures? A prospective observational study. *Epilepsy & Behavior*, *94*, 258–263. https://doi.org/10.1016/j.yebeh.2019.03.034

13 Neuropsychology of Dementia

Introduction

Case Study

"A 37-year-old male patient visited the outpatient clinic with complaints of gradual cognitive decline, which had started three years earlier. Working as an industrial researcher, he started to make serious calculation mistakes that made him quit the job and begin working as a manager in a company. However, his frequent forgetfulness, along with aggravation in recent memory impairments, hampered him from fulfilling his duties, making him change jobs frequently. Apraxia and apathy had started two years before he visited our clinic, and disorientation to time and person worsened to a degree where it became impossible to commute daily between his workplace and home. At the time of his visit to our clinic, not only was he fired from his recent job, but also he needed frequent reminders from his family to maintain hygiene. His sleep disturbance became prominent, frequently waking up in the middle of the night self-talking.

Before his visit to our clinic, he had visited two hospitals for evaluation and management of his symptoms, but to no avail. For a thorough examination of his symptoms, he was immediately admitted to our psychiatric ward. His laboratory findings did not reveal any abnormalities, and his tests for human immunodeficiency virus and syphilis all turned out to be negative. Upon his psychiatric admission, a neuropsychological test battery was implemented to evaluate the patient's cognitive status. He scored 22 in the Mini-mental status examination, one on the Clinical dementia rating scale (CDR), and 4.5 on the Clinical Dementia Rating-Sum of Box score (CDR-SB). In his cognitive tests, in contrast to his relatively preserved language function, he displayed severe impairments in free recall, 20-minute delayed recall, and recognition.

Brain magnetic resonance imaging demonstrated global cerebral atrophy of grade 1 by cortical atrophy scale and notable medial temporal lobe atrophy of grade 2 by medial temporal lobe atrophy visual rating scale. Atypically early onset of dementia symptoms made the patient an eligible candidate for amyloid positron emission tomography (PET) imaging. 18-Florbetaben PET images revealed diffuse amyloid deposition with a score of 3 in brain

DOI: 10.4324/9781032640839-16

beta-amyloid plaque load (BAPL), with predominant amyloid deposition in the striatum. The patient's history, along with neuroimaging results and cognitive test results, all satisfied the National Institute of Neurological and Communicative Disorders and Stroke and the Alzheimer's Disease and Related Disorders Association Alzheimer's (NINCDS-ADRDA) criteria 16 for probable Alzheimer's disease with the high level of evidence. Therefore, 5 mg of donepezil was prescribed, and the patient was discharged on the tenth day of his admission. In control of his persistent cognitive decline post-discharge, the dosage of donepezil was increased to 23 mg with a combination of memantine, which was also increased to 20 mg. His cognitive decline has been relatively plateaued, but we advised the patient and his caregiver to visit the clinic to monitor his symptoms regularly (Um et al., 2017).

The case study above demonstrates most of the signs and symptoms along with the tools and investigation techniques that are typically used in the diagnosis of dementia of Alzheimer's type. As the world in most countries has grown older, many disorders have become prominent as mainstays of the populace. The human body is an incredibly fragile ecosystem, and it takes much effort to maintain its performance and capability.

Dementia is a syndrome that indicates a loss of neurocognitive functions like memory and significant deterioration in the ability to carry out activities of daily living (ADLs). This deterioration is progressive and involves different areas like memory, learning, attention, judgment, and affective changes (Gupta et al., 2011). Of the various components that have been mentioned above, memory is more salient than these. The risk factors that may lead to dementia include age, high blood pressure, high sugar level, excessive drinking or smoking, etc. The disease presents itself in various forms. Some symptoms may include forgetting familiar names and places, memory loss, poor judgment and decision-making skills, poor social cognition, inconsistent changes in behavior and personality, hallucination, and difficulty in interaction with familiar people, places, or things. Presentation of dementia may vary from person to person as it also depends on underlying health conditions, neurological health, and related factors. Dementia can be caused by multiple diseases which impact the brain and cause cognitive and behavioral dysfunction. It affects the psychological, behavioral, and psychosocial health of the person and those around him. According to a recent update from WHO, approximately 55 million people worldwide are affected by dementia, with around 10 million cases yearly. Based on the research until now, there are four different types of dementia: vascular dementia, Lewy body Dementia, frontotemporal dementia, and Alzheimer's disease. There is a fifth category which can be a mixed representation of all these.

Causes and Classification

Vascular

The role of vascular disease in the etiology of dementia is involved in cases where there are direct relationships between cerebrovascular events and the onset of dementia. Onset is likely to be abrupt or follow a rise and plateau pattern. Physical problems include incontinence, low mobility, and balance problems. This dementia type typically means that there is a disease that manifests in cognitive impairment that results from cerebrovascular disease or hemorrhagic brain injury. The terminology of the disorder has been changed to encompass its impact better. Dementia is a small aspect of the disease, and therefore, the inclusive term "vascular cognitive impairment" (VCI) has been recommended as its replacement. There are more overlaps with dementia of the Alzheimer's type in this case, and therefore, there are more similarities than differences between these alleged subtypes of dementia. Regarding the pathophysiology of the VCI, there are common factors with Alzheimer's Dementia. The risk factors involved are the same as those engaged in other cerebrovascular diseases like age, sex, or genetic factors.

Furthermore, there are chances that psychological stress and other life events may also play a part. Good protective or preventive factors include high educational attainment, eating fish, physical exercise, and a multivitamin diet. The stroke had played a significant part in its etiology. Dementia is often a direct consequence of vascular lesions in the brain. Recurrent stroke causes damage to the blood vessels and leads to the white matter lesions that contribute to overall cognitive decline and post-stroke dementia. Genetic factors do play an essential role in the causality of the disease. There are either genes that predispose a person to the disease or determine the response of the body to the disease. There is likely to be a combination of factors that are affected when it comes to the origin factors. There are subtypes of vascular dementia that are based on the etiology as well. Vascular dementia from severe VCI seems to be the result of accumulating lesions along with cerebral atrophy (Iemolo et al., 2009; Kalaria, 2018).

Lewy Body

Lewy body dementia or dementia with Lewy body (DLB) is a disease that is associated with deposits of alpha-synuclein proteins in the brain. These deposits are known as Lewy bodies that affect chemicals in the brain. These lead to problems with movement, behavior, and mood. Lewy bodies are named for D. Friedrich Lewy, who discovered the proteins in 1912 in people with Parkinson's disorder. A healthy brain has these proteins that play various roles, particularly at synapses. When these proteins start to clump at the synapses, they deprive the neurons of the stimulation. This deprivation leads to their eventual damage and death. The major areas that are affected by the disorder are the cerebral cortex, limbic cortex, hippocampus, midbrain, basal ganglia, and brain stem. Features of this type of dementia are fluctuations of awareness in daily life and tremors of movement. Visual hallucinations or delusions are likely to occur. Falls are also likely to occur as a result of the loss of consciousness. The cause of Lewy body dementia is unknown, but there are specific correlations between the accumulation of the Lewy bodies and two crucial neurotransmitters, namely acetylcholine and dopamine. Age is also a risk factor, where anybody above the age of 50 stands a higher risk than others. Other risk factors are associated with comorbidity with Parkinson's disease, REM Sleep behavior disorder, and the male gender. Genetic contributions have been in the form of variants in the APOE (Apolipoprotein E), SNCA, and GBA (Beta-glucocerebrosidase) genes. In terms of incidence and prevalence rates, dementia with Lewy body is the third most common cause of dementia after Alzheimer's dementia and Vascular Dementia. Some of the core features of the disorder are fluctuating attention, recurrent well-formed visual hallucination, and parkinsonian motor symptoms. This mixture of symptoms has also led to LBD or DLB being used as an umbrella term for the presence of Lewy bodies and Parkinson's dementia (Grover & Somani, 2016).

Fronto-Temporal

Frontotemporal dementia is also known as Pick's disease and is characterized by deficits in behavior, language, and executive functions that occur before the memory deficits. There are changes in the manner and behavior. There is also a loss of social awareness and insight that are more common than memory issues. It is also characterized by loss of neurons and microvascular changes in frontal lobes, anterior temporal lobes, anterior cingulate cortex, and insular cortex. There are two variants of FTD, known as behavior variant and primary progressive aphasia. The former features personality changes that lead to interpersonal problems, while the latter accounts for problems with language, speaking, writing, and comprehension. There are three proteins, namely microtubule-associated protein tau, the TAR-DNA protein, and the sarcoma protein. When these are deposited in abnormal amounts, there is a possibility of FTD occurring. There are specific subtypes related to these proteins, as well (Grover & Somani, 2016).

Mixed

The mixture of two or more of the dementias mentioned above in the same person. There are chances that a similar dementia type might be able to exhibit similar features, indicating the absence of rigid boundaries between these subtypes.

Assessments

Assessments and screening are among the first pieces of evidence that is collected as an indication of the diagnosis that will be presented later. These assessments can take many forms, from simple screening tools to neuropsychological batteries and neuroimaging techniques. In this section, we will screen and review some of the more commonly used techniques and methods used for dementia.

There is now greater recognition of the pre-dementia phase in neurocognitive conditions like Alzheimer's disorder. This discovery has increased the research into the biomarkers and signs that might prevent or postpone the onset of dementia. While knowledge of the early phase might better inform the steps that would be taken later, ignorance is bliss. There are chances of false positives that are more likely to harm than be good news for the patients. In these cases, the onus is on the patients to decide on the amount of information that they would like to receive regarding their condition. Mild cognitive impairment (MCI) is characterized by a reduction in cognitive capabilities at a rate that is unexpected for an individual's age and education. However, this decline is not so significant as to interfere with a person's daily life. Early dementia is the onset of a neurodegenerative disorder at a much earlier age than usual (Panegyres et al., 2016).

Screening Tools

Mini-Mental State Exam

The Mini-Mental State Exam (MMSE) is a widely used test of cognitive function among the elderly. The battery includes a test for orientation, attention, memory, language, and visual-spatial skills. Thirty points can be awarded overall, and it measures the full range of cognitive functioning, especially for the elderly. The entire administration of the tests can be done within 5–10 minutes. Any score of 24 or above is an indication of normal cognitive functioning. Less than 9 points on the MMSE is indicative of severe impairment, while 10–18 is moderate, and the rest indicates mild. The main issues with the MMSE are that it is a purely verbal scale, and there is the possibility of the patient not being able to hear the questions rather than not being able to answer them.

Clinical Dementia Rating Scale

The CDR includes semi-structured interviews of the patient and their informant as well as other forms of testing in order to rate the patient on dimensions like memory, orientation, judgment, problem-solving, community affairs, home and hobbies, and personal care. Each dimension is rated on a scale of 0–5, where 0 is none, 0.5 is presence, 1 is mild, 2 is moderate, and 3 is severe. The CDR only rates impairment that results from cognitive impairment. All of these scores are then calculated to derive the global CDR that is similarly rated on a four-point Likert scale. One of the significant limitations of the CDR is that it does not function well as a brief assessment tool. Much of the scoring is dependent on the interviews and clinical observations (Morris, 1997).

Global Deterioration for Aging and Dementia

The Global Deterioration Scale (GDS) is based on a seven-point Likert scale to assess the severity of cognitive and functional impairment in healthy aging, memory impairment with age, and degenerative dementia. The scale is arranged in stages for classifications. The first stage indicates no cognitive decline, where there are no complaints of memory deficits, and the clinical interview does not elicit anything similar. The second stage is very mild cognitive decline, and there are complaints of forgetting about where familiar objects are placed. There are concerns concerning the symptoms, and the patient performs below their expected level on the WAIS vocabulary scale. The third stage is mild cognitive decline, where the clinical symptoms of forgetfulness appear. There is also difficulty in remembering names and reading as well. This decrease in performance becomes evident in social and occupational scenarios as well. The patients are also likely to get lost traveling in an unfamiliar environment. Moderate Cognitive decline is assessed in the fourth stage of the scale. At this stage, the patient has difficulty concentrating, and there is a decrease in the knowledge of current affairs as well as their personal lives. Denial becomes the primary defense mechanism. Orientation remains intact, and there is no difficulty in traveling in familiar areas. Stage 5 is the moderately severe cognitive decline, and the second to last and last stage is severe and very severe cognitive decline. All of these indicate an increasing deterioration in the functionality and performance of everyday activities. The fifth stage marks an end to independent living, where there has to be some assistance with activities of daily living. Disorientation to time and place begins to set in. The level of deficit becomes evident on the mental status questionnaire as well. These dimensions of every life deteriorate as time goes on, and the patient progresses to the next two levels. This tool has many advantages over other rating instruments. It is designed explicitly for progressive primary dementia, and it reflects the stages accurately. It also provides the clinician with enough leeway to incorporate patient characteristics like culture, socioeconomic status, and education to create a more accurate diagnosis and remove biases (Reisberg et al., 1982, 1988).

Montreal Cognitive Assessment

The Montreal Cognitive Assessment (MoCA) is another tool that was specifically developed to detect cases of mild cognitive impairment to ascertain these cases quickly. This attempt is crucial research as it leads to the identification of a clinical stage that eventually proceeds to dementia – the MoCA, which is a 30-point single-page assessment that can be administered in 10 minutes. There are five points for short-term memory recall tasks that have to learn trials of giving nouns and delayed recall. Visuospatial ability is assessed by the clock drawing task and a three-dimensional cube drawing. Executive functioning is measured through trail-making and phonemic fluency, and finally, there is a verbal abstraction task. Attention, concentration, and working memory are assessed using the serial subtraction and digit span tests. Language is evaluated using naming tasks and repetitions of sentences (Nasreddine et al., 2005).

Neuroimaging

The standard protocol for dementia patients is the non-contrast MRI that includes images of the sagittal and axial T1 images, axial T2, FLAIR, and diffusion-weighted images. There is an MRI interpretation that includes evidence of cortical strokes, multiple lacunar, and severe subcortical white matter disease that would be considered to be etiology for dementia. Brain imaging techniques like computed tomography (CT) and magnetic resonance imaging (MRI) would help in diagnosis by allowing the doctors to see the changes in the brain structure and function that would indicate dementia. The scans are more likely to be helpful in cases where there is uncertainty about the type of dementia. This technique would also be useful in cases where there is a combination of dementia present as well. In cases like Alzheimer's disease and vascular dementia, neuropsychological assessments are better suited. These are only useful as corroborating evidence rather than being used as independent diagnosis evidence (Borghesani et al., 2010).

Interventions

Psychoeducation

Ponce et al. (2011) demonstrated that psychoeducational intervention in family caregivers was essential for families that had a person with dementia. One of the primary goals of such psychoeducational interventions is to prepare the family members to monitor the course of the disease and be alert to the possibility of a relapse in symptoms. Group sessions can also be included so that there is no feeling that the family is struggling alone. Many people are often going through the same thing, but in isolation, it is easy to forget that. Depending on the group dynamics, certain activities like video debates on Alzheimer's disease, discussion on articles and documentaries, open dialogue, expressing emotions, and problem-solving for everyday difficulties would be conducted. More often than not, the caregivers tended to be females, residing with a person with Alzheimer's and would be a wife or a daughter. The psychoeducational intervention led to the caregivers reporting less burden and showed more considerable improvement in terms of general strain, isolation, disappointment, emotional involvement, and their environment. The caregivers play a vital role in the prognosis of dementia to the exclusion of their own lives. Therefore, it was identified that a multi-disciplinary team be assigned for them as well. Lewis et al. (2010) developed a module for internet-based dementia caregivers. This psychoeducational program provides the caregivers with the knowledge, skills, and outlook necessary to succeed. The program that was developed was able to provide expert as well as peer information. This intervention led to an increase in knowledge that also facilitated increasing self-efficacy and confidence in their ability to fulfill their roles adequately.

Psychosocial Rehabilitation

Dementia is at par with disabilities that lead to issues in dealing with everyday lives. A rehabilitation-centered approach is more likely to focus on improving the standard of living and increasing the level of access for people with dementia. In this manner, it is person-centered and includes an individualized formulation that discards the one-size-fits-all policy that other interventions tend to follow. There are many types of intervention that do not strictly fit the criteria of cognitive rehabilitation. Some are aimed at self-management or self-efficacy enhancement.

Along with these, some are relevant to positive self-growth and enjoying pleasurable activities. The rehabilitation model is likely to offer a tremendous opportunity that would not be possible through other forms of interventions. It is a more practical means of providing person-focused, evidence-based practices (Clare, 2017). According to a review of the various models in the last decade, there was a significant improvement in physical interventions, where there was an improvement in physical and cognitive functions along with ADL skills, but no improvement in mood or behavioral outcomes. Cognitive stimulation has been proven to be effective, mainly when conducted in a group. There was difficulty in differentiating between psychosocial and purely psychological interventions. There were ones that included psychotherapeutic approaches like cognitive-behavior therapy (CBT). Hence, these were considered in one group itself – these addressed components like staff training and carer perceptions of the problematic symptoms of the person with dementia. Social integration of people with dementia is more likely to be beneficial as opposed to individual therapy or isolated models of treatment (McDermott et al., 2019).

Prevention

Dementia of all types can only be managed, and to some extent, its progress is halted or delayed. There are no cures for it yet, and the functionality that is lost is permanent in most cases of true dementia. Prevention of this condition would include all methods that are performed to eradicate, eliminate, or minimize its impact. Timing is most important when it comes to the prevention of neurodegenerative disorders. This issue becomes more prevalent as the beginning of the disorder is much earlier than the physical manifestations that alert the individual to the possibility of the disorder or disease in the body. Physical activity and exercise have been linked with the reduced risk of many disorders, not just neurodegenerative ones. As the number of activities increased, there was a proportional reduction in the risk of Alzheimer's. Along with physical activities, it is also essential that intellectual activities are also maintained. There are chances that these act as exercises for the brain and builds up the cognitive reserve. The cognitive reserve is a stock of brain function and intellectual progress that delays the impact of neurodegenerative diseases. Education and cognitive activities are more likely to enable people to use better compensatory methods to nullify the impact of dementia. A Mediterranean diet has also proven benefits. There are nutritional and behavioral recommendations that go together. Physical activity and fresh fruits and even wine, fish, red meat, and eggs are included in this diet. Lastly, the maintenance of social networks and regular social interaction would lead to better management of the decline in cognitive functionality (Savica & Petersen, 2011).

Research Perspective

In recent years, researchers have focused on finding the prognostic markers, biomarkers for different types of dementia, early markers, and markers of progression and on improving precision in diagnosis for an early diagnosis of the diseases. They have been studying individual cases, cross-cultural effects, and disease progression under different conditions. With the sharp increase in population and deteriorating health conditions, cases of dementia are also rapidly rising. While this is happening, the general population is still ill-informed about dementia. The most common challenges observed worldwide are paucity of resources and funding for research, lack of advanced neuroimaging methods for diagnosis and assessment of disease, and scarcity of resources to manage patients with dementia. An extensive analysis of the Global Dementia Prevention program in countries like India, China, Nigeria, and Iran has indicated a paucity of culturally appropriate assessment tools and advanced neuroimaging methods (Chan et al., 2019).

Advancement in knowledge about dementia can be better understood in four different disease progressions. The first step is to analyze and manage the risk, and the second step is to accurately pre-diagnosis; the third step is the correct diagnosis of the disease using the information on cognitive, behavioral, and psychosocial symptoms; and the fourth and the last one is to devise a plan for management keeping in mind the patient, their caregivers, and their immediate environment.

In another review by Franzen et al. (2020), where they reviewed dementia assessment tools in non-western countries and low-educated or illiterate populations, they found gaps in the standardization of neuropsychological tests available for the assessment of cognitive ability. Due to the differences in language, culture, and social conditions, some of the tests developed in Western countries are not applicable precisely as they are. The inconsistency in adaptation procedures for using the tests in new settings and the use of both neuropsychological tests and neuroimaging methods for assessment by a smaller fraction of studies were found. These indicate a broad scope of research for developing assessment tools in cross-cultural settings, studies with a large sample for the validity of tests for different cognitive domains. It is essential to keep in mind that changes in language and education play a crucial role in performance in neurocognitive tests, and this remains an area with immense opportunity for research.

Vascular dementia is caused by a disruption in cerebrovascular blood flow due to certain medical conditions which impact brain health. This results in cognitive impairments leading to dementia. For example, impairments in executive function, visuospatial abilities, language processing, remembering, etc. Recognition and management of vascular cognitive impairment and dementia (VCID) is complex primarily because of the overlap of symptoms with age-related cognitive decline and second, because it requires symptomatic identification and treatment plan. Researchers are working towards finding stable and specific biomarkers to identify the dysfunctions in autonomic regulation in a way to diagnose dementia at earlier stages. Neuroimaging biomarkers can prove to be critical for diagnosis as they can help in the identification of pharmacological causes before symptomatic severity. A crucial area of study is in neuroimaging methods like diffusion tensor imaging, amyloid-b-PET imaging, along with other methods for unique analysis and quicker assessment (Bir et al., 2021; Jagtap et al., 2015).

Behavioral variants of frontotemporal dementia (bvFTD) and Alzheimer's disease have overlapping cognitive symptoms, which makes it harder for accurate assessment and management plan on time. Practitioners in the field can benefit from molecular biomarkers and neuroimaging methods for assessment in the early stages of the disease. Machine learning-based models are being tested for precise diagnosis using cognitive tests (Garcia-Gutierrez et al., 2022). Onset symptoms of BvFTD are similar to many psychiatric disorders, such as personality disorders, schizophrenia, Autism spectrum disorder, etc. Researchers are proposing using social cognition tests for assessment in neuropsychological batteries, genetic testing for mutation to identify the onset of the disease and clinical differential diagnostic methods for better and early diagnosis of the disease (Ducharme et al., 2020). Lewy body dementia has a variety of symptoms like motor, autonomic, neuropsychiatric, and cognitive, and their presentations vary from person to person, changing with time. Lewy body dementia (LBD) requires the management of both neurogenerative and non-neurogenerative symptoms. There is a need for a multi-disciplinary management and treatment approach for LBD focusing on symptoms as presented by each patient (Taylor et al., 2020).

Since it has been observed that good physical and brain health reduces the risk of dementia, there has been an increase in studying risk reduction and management methods as well. The primary risk factors for dementia include excessive weight gain or weight loss, major depressive disorder, bipolar disorder, schizophrenia, excessive alcohol intake, and unhealthy lifestyle. Psychological and pharmacological intervention methods to address lifestyle changes and personalized risks related to health conditions are emerging research areas. While developing any diagnostic tool or method of assessment or intervention, the most common challenges faced are small sample size, lack of appropriate control (on the extraneous variables), lack of long-term efficacy of new interventions, etc. There is a dire need for longitudinal studies to show the long-term effects of pharmacological interventions and to have more clinical trials to test more effective clinical measures to manage non-degenerative symptoms of dementia.

References

Bir, S. C., Khan, M. W., Javalkar, V., Toledo, E. G., & Kelley, R. E. (2021). Emerging concepts in vascular dementia: A review. *Journal of Stroke and Cerebrovascular Diseases, 30*(8), 105864. https://doi.org/10.1016/j.jstrokecerebrovasdis.2021.105864

Borghesani, P. R., DeMers, S. M., Manchanda, V., Pruthi, S., Lewis, D. H., & Borson, S. (2010). Neuroimaging in the clinical diagnosis of dementia: Observations from a memory disorders clinic. *Journal of the American Geriatrics Society, 58*(8), 1453–1458.

Chan, K. Y., Adeloye, D., Asante, K. P., Calia, C., Campbell, H., Danso, S. O., Juvekar, S., Luz, S., Mohan, D., & Muniz-Terrera, G. (2019). Tackling dementia globally: The global dementia prevention program (GloDePP) collaboration. *Journal of Global Health, 9*(2), 1–5.

Clare, L. (2017). Rehabilitation for people living with dementia: A practical framework of positive support. *PLoS Medicine, 14*(3), e1002245.

Ducharme, S., Dols, A., Laforce, R., Devenney, E., Kumfor, F., van den Stock, J., Dallaire-Théroux, C., Seelaar, H., Gossink, F., Vijverberg, E., Huey, E., Vandenbulcke, M., Masellis, M., Trieu, C., Onyike, C., Caramelli, P., de Souza, L. C., Santillo, A., Waldö, M. L., … Pijnenburg, Y. (2020). Recommendations to distinguish behavioural variant frontotemporal dementia from psychiatric disorders. *Brain, 143*(6), 1632–1650. https://doi.org/10.1093/brain/awaa018

Franzen, S., Van Den Berg, E., Goudsmit, M., Jurgens, C. K., Van De Wiel, L., Kalkisim, Y., Uysal-Bozkir, Ö., Ayhan, Y., Nielsen, T. R., & Papma, J. M. (2020). A Systematic review of neuropsychological tests for the assessment of dementia in non-western, low-educated or illiterate populations. *Journal of the International Neuropsychological Society, 26*(3), 331–351. https://doi.org/10.1017/S1355617719000894

Garcia-Gutierrez, F., Delgado-Alvarez, A., Delgado-Alonso, C., Díaz-Álvarez, J., Pytel, V., Valles-Salgado, M., Gil, M. J., Hernández-Lorenzo, L., Matías-Guiu, J., Ayala, J. L., & Matias-Guiu, J. A. (2022). Diagnosis of Alzheimer's disease and behavioural variant frontotemporal dementia with machine learning-aided neuropsychological assessment using feature engineering and genetic algorithms. *International Journal of Geriatric Psychiatry, 37*(2). https://doi.org/10.1002/gps.5667

Grover, S., & Somani, A. (2016). Etiologies and risk factors for dementia. *Journal of Geriatric Mental Health, 3*(2), 100.

Gupta, R., Sharma, V., & Malhotra, S. (2011). Neuropsychological assessment and management of patients with dementia: A practical approach. In V. Sharma & S. Malhotra (Eds.), *Clinical neuropcycholohy* (pp. 251–263). Harprasad Institute of Behavioral Science.

Iemolo, F., Duro, G., Rizzo, C., Castiglia, L., Hachinski, V., & Caruso, C. (2009). Pathophysiology of vascular dementia. *Immunity & Ageing, 6*, 1–9.

Jagtap, A., Gawande, S., & Sharma, S. (2015). Biomarkers in vascular dementia: A recent update. *Biomarkers and Genomic Medicine, 7*(2), 43–56. https://doi.org/10.1016/j.bgm.2014.11.001

Kalaria, R. N. (2018). The pathology and pathophysiology of vascular dementia. *Neuropharmacology, 134,* 226–239.

Lewis, M. L., Hobday, J. V., & Hepburn, K. W. (2010). Internet-based program for dementia caregivers. *American Journal of Alzheimer's Disease & Other Dementias®, 25*(8), 674–679.

McDermott, O., Charlesworth, G., Hogervorst, E., Stoner, C., Moniz-Cook, E., Spector, A., Csipke, E., & Orrell, M. (2019). Psychosocial interventions for people with dementia: A synthesis of systematic reviews. *Aging & Mental Health, 23*(4), 393–403. https://doi.org/10.1080/13607863.2017.1423031

Morris, J. C. (1997). Clinical dementia rating: A reliable and valid diagnostic and staging measure for dementia of the Alzheimer type. *International Psychogeriatrics, 9*(S1), 173–176.

Nasreddine, Z. S., Phillips, N. A., Bédirian, V., Charbonneau, S., Whitehead, V., Collin, I., Cummings, J. L., & Chertkow, H. (2005). The Montreal Cognitive Assessment, MoCA: A brief screening tool for mild cognitive impairment. *Journal of the American Geriatrics Society, 53*(4), 695–699.

Panegyres, P., Berry, R., & Burchell, J. (2016). Early dementia screening. *Diagnostics, 6*(1), 6. https://doi.org/10.3390/diagnostics6010006

Ponce, C. C., Ordonez, T. N., Lima-Silva, T. B., Santos, G. D. dos, Viola, L. de F., Nunes, P. V., Forlenza, O. V., & Cachioni, M. (2011). Effects of a psychoeducational intervention in family caregivers of people with Alzheimer's disease. *Dementia & Neuropsychologia, 5,* 226–237.

Reisberg, B., Ferris, S. H., de Leon, M. J., & Crook, T. (1982). The Global Deterioration Scale for assessment of primary degenerative dementia. *The American Journal of Psychiatry, 139*(9), 1136–1139.

Reisberg, B., Ferris, S., De Leon, M., & Crook, T. (1988). Global deterioration scale (GDS) Psychopharmacol. *Bull, 24*(4), 661–663.

Savica, R., & Petersen, R. C. (2011). Prevention of dementia. *Psychiatric Clinics, 34*(1), 127–145.

Taylor, J.-P., McKeith, I. G., Burn, D. J., Boeve, B. F., Weintraub, D., Bamford, C., Allan, L. M., Thomas, A. J., & O'Brien, J. T. (2020). New evidence on the management of Lewy body dementia. *The Lancet Neurology, 19*(2), 157–169. https://doi.org/10.1016/S1474-4422(19)30153-X

Um, Y. H., Choi, W. H., Jung, W. S., Park, Y. H., Lee, C.-U., & Lim, H. K. (2017). A case report of a 37-year-old Alzheimer's disease patient with prominent striatum amyloid retention. *Psychiatry Investigation, 14*(4), 521.

14 Alzheimer's Disease

Introduction

The previous chapter discussed dementia and its various subtypes, but there was one that was conspicuous in its absence. Alzheimer's disease (AD) or dementia is one of the most widely researched and published forms of dementia. This distinction alone warrants an entire chapter dedicated to the disease that often robs a person of their true selves. AD has become so infamous that it has become synonymous with the term dementia itself. However, there are many types of dementia, and these are not always the inevitable consequence of aging either. Three criteria are used to differentiate between the various types and subtypes of dementia. The first criterion is the location, the corticals, or the subcorticales. Cortical dementias are those that start by affecting the grey matter. AD is included in this category of dementias. The subcortical dementias are a disease that affects the white matter. These are the wires or the connections between the various areas of grey matter. However, this criterion is not strictly adhered to as dementia themselves do not conform to these boundaries. Nowhere is this more apparent than in the case of AD, which, while being classified as cortical dementia, also affects the hippocampus, which is a subcortical limbic structure. Loss of autobiographical memory, which is a prominent feature of AD, is a result of this effect. The second criterion is the nature of dementia, whether its effects remain static or they tend to get worse with time progressively. Diseases like AD, Pick's, and Huntington's are the ones that follow a general pattern of decline in cognitive and behavioral functions.

In contrast, there are other substances, mostly neurotoxic ones that lead to a static state of dementia, which usually resolves itself once the substance is removed. Irrespective of their nature, there are still differences within the criterion in terms of the rate at which the symptoms tend to manifest. The final criterion is the reversibility of the condition. Most of the research has tended to focus on the progressive and irreversible forms of dementia, but some conditions mimic the dementia conditions but are also usually easy to reverse. This criterion could also be attributed to the problem of misdiagnosis, where the condition could be delirium rather than true dementia (Zillmer

DOI: 10.4324/9781032640839-17

& Spiers, 2001). While there are more research possibilities now than ever, there are no proven fixes that could reverse the brain damage that dementia causes. Neurons are delicate at the best of times, and there would have to be extreme care taken even when these are repaired. The rest of this chapter will focus on the history of AD, starting from Alois Alzheimer's to exploring the current state of research in terms of reversibility and management. Methods of assessment and indigenous practices in India regarding AD will also form a separate section. Finally, methods of prevention and risk reduction for AD would be deliberated.

History

In the year 1906, on November 3, Alois Alzheimer, a clinical psychiatrist and neuroanatomist by profession, reported the presence of severe disease in the cerebral cortex. The date was the 37th Meeting of the South-West German Psychiatrists. Alois Alzheimer was a lecturer at Munich University at the time. The female patient was named Auguste D. and had been under his observation since the year 1901 when he had been a senior assistant at Frankfurt Psychiatric Hospital. The case was brought up by the husband when he noticed his 50-year-old wife was becoming increasingly paranoid and would suffer from bouts of sleeplessness, disturbances in memory, aggressiveness, crying, and progressive confusion. The woman was in the inpatient setup until her death in 1906. During her autopsy, Alzheimer was able to investigate her brain in order to figure out the exact cause of the disease. It was during this that Alzheimer discovered the histological alterations that would later be known as plaques and neurofibrillary tangles. When he presented these findings to Emile Kraepelin, he was encouraged to present these findings as soon as possible. At the time, these histological findings were not connected with any existing clinical symptomatology. Between the years 1906 and 1909, Kraepelin prepared the 8th edition of his famous textbook *Psychiatrie A*. He termed the case of Auguste D. as suffering from AD. Kraepelin's authority for the case made AD the diagnostic term. However, the disease was relatively rare, and Alois Alzheimer's was nearly forgotten for the better part of almost 50 years after its discovery (Hippius & Neundörfer, 2022).

The times have certainly changed since then. Alzheimer's has become a common household name, and many older adults stand at risk of suffering from AD. The amount of research that has gone into the disease is a direct result of its increasing prevalence and incidence rates around the world. Mathuranath et al. (2012) published a report on the incidence rate of AD in India that focused on the Indian state of Kerala. Of the 1,066 participants who were followed during this time, nearly 104 of them developed AD. It can be extrapolated to an incidence rate of 11.67%, where there are many people above the age of 65. The incidence rate of AD increased significantly and in proportion to the increasing age.

Etiology

Genetic

The first hint is that there would be a genetic component to the disease. People who had Down's syndrome would eventually end up suffering from AD as well if they survived till middle age. Although genes are not in complete control of the disease, it does shed considerable light on the nature of the disease. In particular, it is the gene that controls the amyloid-β proteins that tend to accumulate in and around the neurons themselves. Along with the amyloid proteins, the tau protein has also been implicated in the disease. These proteins are instrumental in the intracellular support structure of the axons. These proteins tend to work in tandem and enhance the damage that each does individually (Kalat, 2015). There are many sources on the Internet related to AD. One of the more reliable sources is the website of the National Institute on Aging, which is also part of the US Department of Health and Human Services. There are chances that family history is an important cause, but it does not guarantee that you will have it. It just increases the chances that it develops. Before we delve further into the genetics behind Alzheimer's, it is crucial to understand the differences between the two types. These are differentiated based on the timing of onset: late and early.

The late onset starts in the mid-60s and involves the apolipoprotein-E (APOE) gene. It is also the standard type of the disease, while early onset is rare. This type appears in the person's mid-30s to mid-60s and they are caused by gene changes passed from parent to child (*What Causes Alzheimer's Disease? National Institute on Aging*, 2019). All genes are expressed in terms of different forms known as alleles. The APOE gene also comes in three different types, and there are two APOE alleles from each parent. APOE ε2 allele is relatively protective against the disorder and delays the onset of the disease, than if a person had the APOE ε4 gene. APOE ε3 allele is the most common form that is also neutral in its contribution to the disease. The APOE ε4 gene increases the risk for the disease and having two of these from the biological parent makes the risk of developing the disease profound. About 25% of people carry a single copy of this, and about 2–3% carry both copies (*Alzheimer's Disease Genetics Fact Sheet*, 2019).

Psychological and Behavioral

Goate et al. (1989) represent some of the earlier research on the subject of the genetic etiology of AD. An essential genetic link is a positive family history of AD. That being said, this excludes many cases where there is no history and yet be present in the offspring. There may be two reasons for this exception as well. The etiology of AD is heterogeneous enough to warrant a look at factors apart from genetics. The second reason is that AD only rarely expresses itself before the age of mid-30. There is a likelihood that the people who are predisposed to AD die due to other causes as well. The genetic component is more probable in cases where there is an early onset that is a rare entity. The genetic hypothesis for AD includes three types where the duplication of the amyloid gene causes sporadic AD; familial AD leads to over-expression. There is also the presence of AD-like pathological symptoms in Down's syndrome, where there are three copies of the amyloid gene. Apart from the risk factor that was associated with the APOE ε4 gene, there are also the APP, PSEN1, and PSEN2. These are the amyloid-β precursor protein, Presenilin 1 and Presenilin 2. Genetic linkages studies are the most powerful tool when it comes to the study of Mendelian diseases such as AD. These detect the chromosomal location of the disease genes and lead to the observation that links the genes that are physically close to those and are likely to remain together during the meiosis process. This traditional process has now been substantiated by the development of genome-wide genotyping and second-generation sequencing technologies. These do not identify the genes but rather the loci that could contribute to the development of AD. These loci, in some cases, are part of various biological pathways as well, thus expanding the effect of these processes. The three main pathways that were identified as a result are the immune system and inflammatory responses; cholesterol and lipid metabolism; and endosomal vesicle recycling. The significant advantage of this process is that the innate immune system and the microglia activation/inflammation along with the brain cholesterol metabolism give new methods to control the march of AD (Guerreiro & Hardy, 2014). Genetic variation has been implicated in the functions of the immune systems and lipid metabolism as well. Such genetic data reveals high susceptibility to illness. These also lead to better and more positive outcomes in the long run (Jones et al., 2010).

Down's syndrome has been implicated as a precursor of sorts that might be indicative of AD. There are similarities between these that are not easily ignored. Both disorders have a strong basis in genetic etiology and often manifest in severe behavioral and psychological symptoms. The behavioral and psychological symptoms of dementia (BPSD) is a psychiatric entity that has avoided research due to the heterogeneity in the Down's syndrome cohorts. This research would have far-reaching consequences in the prevention of risk factors for Alzheimer's so that its management begins at an earlier period. Even Alois Alzheimer described the presence of neuropsychiatric symptoms like hallucinations, delusions, paranoia, and agitation in his case. However, these are not given the due consideration that is necessary. BPSD contributes to a reduced quality of life, increased risk of mortality, and a severe burden on caregivers and relatives. There was also an emphasis on the differentiation between psychiatric symptoms and behaviors that would otherwise be a natural consequence of suffering from the illness. There were reports of prolonged periods of inactivity in people with DS, which was nothing but a result of constant hospitalizations. The BPSD runs the spectrum from activity to affective disturbances like apathy, isolation, and depression. Apart from these, agitation, aggressiveness and anxiety, psychosis, and diurnal rhythm disruptions are also associated. All of these symptoms are easier when they are classified into symptom groups such as affective, psychosis, hyperactivity, and euphoria. While these symptoms can be studied in isolation, this takes them away from the other symptoms of the syndrome. While this does not necessitate all the symptoms being studied together, there would be more significant benefits and better outcomes in the knowledge that there are many interrelations between these disorders. These would inform the types of interventions to be used and also account for the comorbidity as well (Dekker et al., 2015; van der Linde et al., 2014).

Assessment

Sheehan (2012) compiled a list of all the assessment scales that are typically used for diagnosing dementia. The ideal properties of an assessment scale would include face reliability and construct and concurrent validity. Also, the inter-rater and test-retest reliability shows that the scale is consistent and does not deviate from the norms that have been set for the same. Finally, the scale should support practical usage and not strain the patient or the clinician using the tool. When it comes to dementia, cognitive functions are considered vital to the prognosis of the disorder. Some of the tools that are used are the Mini-Mental State Examination (MMSE), the clock drawing test, the Addenbrookes Cognitive Assessment (ACE), and the Montreal Cognitive Assessment (MoCA). These short scales can be administered quickly and are reliable as well. The Alzheimer's Disease Assessment Scale – Cognitive Section (Rosen et al., 1984) is a detailed assessment that takes about 40 minutes to administer. The scale has good sensitivity to change and covers all the major cognitive areas and dementia.

The National Institute on Aging has the Alzheimer's Association developed guidelines for the neuropathologic assessment of AD. These represent a revision that has stood the test of time since the year 1997. The significant difference is that the 1997 criteria required a history of dementia that addresses the question of whether it would be influenced by AD. For the neuropathologic diagnosis, neurofibrillary tangles (NFTs) and senile plaques are essential. The NFTs are composed of abnormal tau proteins and may be present both inside and outside the neurons. These also developed in stages from being restricted to the entorhinal cortex and then its eventual proliferation to the hippocampus and the amygdala. The other major component of this is the senile plaques. These are extracellular deposits of the Aβ peptides. These are more likely to form neuritic plaques more closely associated with a neuronal injury. These plaques are directly responsible for disrupting synaptic communication and glial activation. The presence of these two neuropathological signs is highly correlated with the clinical symptoms of AD as well. Genetic and biomarkers are to be used in various settings so that these aid in the post-mortem diagnosis of AD (Hyman et al., 2012).

Behavioral Pathology in Alzheimer's Disease Rating Scale

This scale started development in 1987, in the context of many similar scales, such as the Symptoms of Psychosis in AD and others. Nevertheless, these often tended to mix up cognitive and behavioral symptoms and also included impairments in function. The Behavioral Pathology in Alzheimer's Disease Rating Scale (BEHAVE-AD) is one of the first that was focused on AD. There was a dedicated effort to identify behavioral disturbances in patients with AD in order to develop a tool that was sensitive to the behavioral symptoms. The BEHAVE-AD is a 25-item scale that measures behavioral disturbances on a 4-point scale of severity. These symptoms are arranged across seven categories. There is also a global scale for the assessment of overall problems. The seven categories are paranoid and delusional ideation, hallucinations, activity disturbances, aggressiveness, diurnal rhythm, affective disturbances, and finally, anxiety and phobias. The scale also enjoys excellent reliability, which is comparable to the MMSE. Its construct validity has been supported by the differences between the various symptoms of AD (Reisberg et al., 1997).

Neuropsychiatric Inventory

Lai (2014) conducted a review article on the merits and problems that are associated with the Neuropsychiatric Inventory (NPI). It was developed by Cummings et al. (1994) as a measure to assess dementia-related behavioral symptoms. This scale also expanded on the various subdomains to 10, which included euphoria, apathy, and disinhibition and also later expanded to add eating abnormalities as well. These new domains ensure that the tool can be used to screen multiple types of dementia that are not limited to AD. The tool is administered to the caregivers of dementia patients. The rating is across a 4-point scale for frequency and a 3-point scale for the severity of the symptoms. The distress caused by the same is recorded on a 5-point scale. The tool enjoys excellent psychometric properties and has reasonably good content validity and inter-rater reliability. Some of its merits are that the tool is comprehensive and yet avoids overlap in symptoms. It is easy to use and flexible to administer. The main problems are associated with the scoring aspect, where there are no odd scales that causes it to lose statistical rigor.

Neurobehavioral Rating Scale

Neurobehavioral Rating Scale (NRS) is a scale that contains 28 items that measures the severity of a broad range of cognitive and non-cognitive symptoms. Each of these items is scored on a scale of 0 (not present) to 6 (extremely severe). This assessment is done based on a structured interview with the patients that lasts about 45 minutes. The items listed range from inattention, expressive deficit, and inaccurate insight to fatigability, blunted affect, poor planning, and fluent aphasia. It enjoys excellent inter-rater reliability and validity. Some results show that the NRS is a useful instrument that can be a comprehensive assessment of cognitive deficits, psychiatric symptoms, and behavioral disturbances that accompany neuropsychiatric syndromes. The use of a semi-structured interview opens up the assessment of a broad range of clinical symptoms (Sultzer et al., 1995).

Brain Imaging in AD

There is a hypothesis that attempts to study whether glucose metabolism and medial temporal lobe brain volumes could be associated with the predictive cognitive decline within the older adult population. Positron emission tomography (PET), structural magnetic resonance imaging (MRI), and functional MRI have all been used to detect any alterations that may occur in the brains of individuals that are at risk of AD. These are all aimed at specific areas that are likely to be at high risk for symptoms of AD. The medial temporal lobe is one such area, as it is the seat of learning and memory. These cognitive functions are the worst affected in AD. These are primarily due to the hippocampus and the entorhinal cortex of the medial temporal lobe. Other conventional approaches use mild cognitive impairment (MCI) as a precursor as it is associated with smaller brain volumes in the structures previously mentioned.

Consequently, brain glucose metabolism in these areas is also below average. PET and MRI images were used to assess the structural and functional alterations in the brain in people that were beginning to show the first signs of AD. The declining rates of temporoparietal metabolism were predictive of the noticeable decline in cognitive functions as evidenced by a behavioral tool like a modified MMSE. The study provides conclusive evidence that metabolic and structural brain alterations predict cognitive issues in healthy older adults (Jagust et al., 2006).

No more significant sign heralds the arrival of AD than the presence of amyloid plaques. Nordberg (2008) provided a PET scanning that specifically targeted amyloid formations. The imaging ligand for amyloids would be useful in identifying and tracking its progress as well. Ligands are chemicals that excel at binding to particular proteins in some areas of the brain. One of the novel forms of intervention that shall be discussed in the next section is anti-amyloid therapy. These are aimed at reducing the strain caused by the plaques that are composed of the amyloid-β. This view was also supported in a review where there were amyloid tracers that were developed to be used with PET. These are currently undergoing clinical trials to assess their levels of efficiency in detecting fibrillar amyloid tangles. These can also be used as biomarkers that can be used to predict the development of AD before the onset of dementia. These tracers are instrumental in the development of newer forms of therapeutic intervention as well (Herholz & Ebmeier, 2011).

Interventions

Anti-Amyloid Therapy

The commonly accepted hypothesis is that the formation of soluble neurotoxic oligomers triggers AD pathogenesis. It occurs from physiologically monomeric Aβ followed by the generation of insoluble polymeric Aβ aggregates that are eventually accumulated as amyloid plaques. Prevention of this cerebral amyloidogenesis via inhibiting pathological Aβ oligomerization is considered the most promising solution to the AD problem. While the main driving forces behind the disorder are unclear, zinc ions play an essential role. The amyloid matrix is an essential aspect of anti-amyloid therapy and leads to the transformation of the neurotoxic oligomers and amyloid plaques (Kozin et al., 2018). Technology has become pervasive in most aspects of our lives; hence it is no surprise that it has also made its mark when it comes to the treatment of AD as well. Here, the focus remains on the amyloid and tau proteins and involves systems modeling to handle these better. Instances of immunotherapy have been shown to reduce tau phosphorylation. This model laid the foundation for the potential of immunity against amyloid-β as well. The use of activated microglial against the amyloid plaques was a novel approach that pitted the two pathologies of AD against each other. The mathematical model was run on computer simulations and may provide one of the better chances to treat AD as more and more mechanisms of the disorder become apparent with time (Proctor et al., 2013).

Psychological Interventions

In order to decide where people were more likely to seek help in the event of a family member dealing with AD, the study sought to understand their beliefs about the several sources of help and their perception of how good they were. Some of the sources that were identified were psychologists, self-help groups, close family members, neurologists, friends, psychiatrists, and religious leaders. On the flip side, there was an awareness of the harm that people like faith healers, sorcerers, and pharmacists could pose. Psychologists were the first choice when it came to asking for professional help. This choice was peculiar as the professionals trained to deal with AD were the neurologists and psychiatrists who were chosen after the psychologists. An explanation for this was that the other professions are pigeonholed in their public perception. Psychiatrists are considered for significant disorders that do not necessarily have anything to do with memory symptoms that are prevalent in AD. There are also chances that general doctors are consulted in cases where there are more behavioral than cognitive issues. When it came to the choice of intervention, most preferred lifestyle changes like eating better, exercising, going to church, and psychotherapy. These are interventions that are viewed less negatively than psychiatric interventions or clinical procedures. Alternative interventions were more likely to be opted for as these are healthy and do not carry the risk of side effects. However, many of them, like taking vitamins or going on vacations, are also the least helpful when it comes to delaying the symptoms of cognitive decline (Blay et al., 2008).

Although the amyloid hypothesis is widely accepted, there are also other hypotheses, notable among these being the cholinergic hypothesis. This hypothesis directly ties in with the use of cholinesterase inhibitors prescribed for dementia. The cholinergic system at the synaptic overlap of neurons may be dysfunctional leading to memory loss and cognitive impairment. Thus, the rejuvenation of the cholinergic system is directly linked with the reduction in the risk of cognitive dysfunction and improvement in the memory of the patient. Acetylcholinesterase is the crucial enzyme in this system and is also directly responsible for the movement of the neurotransmitters. Diagnosis is still dependent on the evaluation of mental status and cognitive testing. The other behavioral symptoms only appear in the advanced stages of AD when it is too late to do anything (Rajasekhar & Govindaraju, 2018).

Prevention – Indigenous Methods

There are myriad ways that have been looked into to treat AD. Treatment of the symptoms has done very little to delay the progress of the disease. The behavioral symptoms are much more easily managed than the neuropathological ones. Some of the prominent risk factors associated with AD are age, gender, family history, APOE ε4, systolic blood pressure, BMI, total cholesterol levels, and physical activity would lead to the calculation of the total risk that is associated with the disorder. The only viable option that leads to successful prevention of the disease would occur in the preclinical stage. The factors mentioned above would be a precursor that highlights the level of risk and the appropriate strategies to be employed in the event. The gradation of the risk factors would view age at a basic level that then carries on to the family history of AD in first-degree relatives divided in severity based on the early and late onset. Finally, the family history combined with the presence of the APOE ε4 and a positive biomarker like the PET amyloid scan. One of the most important non-pharmacological ones is lifestyle changes to modify vascular risk factors like high systolic blood pressure and enhance protective factors through physical exercises and cognitive stimulation, and a healthy diet (Mediterranean diet). When it comes to pharmacological approaches, there are more likely to be cases where the treatment target could either be the symptoms or the neuro pathophysiological attributes of the disease. Some medicines are aimed at reducing the pathophysiology that is most commonly attributed to AD. These are in the form of amyloid deposition, tau hyperphosphorylation, microglial activation, and poor synaptic plasticity (Mathuranath et al., 2012).

Research Perspective

AD is the most widespread and critical kind of dementia. AD follows a multi-factorial disease progression causing cognitive, physical, emotional, and social dysfunction in the aging population. AD deviates from normal aging as it adversely affects the functioning and protection system of neurons in the brain. Over the years, researchers have been constantly aiming at finding pharmacological and non-pharmacological methods to diagnose and treat AD. Since the effects of pathological changes in the brain are visible only after a few years of the actual onset of disease in the form of cognitive and behavioral impairment, furthermore, it causes irreversible biochemical and anatomical changes in the brain, which has made it impossible to find the cure for AD so far. To overcome this hindrance in recent years, researchers have broadened their approach by integrating risk management, biomarker identification, and management techniques for AD. Studies have found that both personal and environmental risk factors can be improved to have healthy aging and avoid or delay the onset of AD. These factors include exercise, taking nutritious food, having a good social life, meditating, living a stress-free life, avoiding excessive drinking or smoking, or intake of any toxic elements. Risk factor prevention is a developing area of research as its impact on the brain is still being studied (Guzman-Martinez et al., 2021). Along with these, researchers have been focusing on finding reliable biomarkers, which include protein biomarkers (amyloid-β and tau protein), cerebral fluid biomarkers, blood biomarkers, and imaging biomarkers, that is, PET scans for amyloid-β and tau protein. These biomarkers will help in the early detection of the disease as well as in designing and management of treatment plans for AD. They have also identified preventive approaches such as nutraceuticals which are multifaceted therapies to control tau protein degradation, mediation, and a variety of exercises, including aerobic exercises, which help in building healthy lifestyles and healthy aging.

Biomarkers are essential not only to diagnose preclinical AD but also to differentiate symptoms from cognitive changes associated with normal aging and from differentiating from other kinds of dementia. In an extensive meta-analysis to find the association between neurocognitive abilities and biomarkers in AD, researchers found exciting results. They found out that differences in neurocognitive performance in older adults may arise earlier in the course of AD than previously thought and are found to be connected with the accumulation of tau protein and amyloid protein retention (Duke Han et al., 2017). They have also proposed cognitive biomarkers to be included in the available biomarkers for the preclinical detection of AD. Cognitive biomarkers may help in the early detection of functional changes in the brain, which may occur before structural changes in the brain as observed by other biomarkers or may be able to detect differential changes than those studied by other biomarkers. Tahami Monfared et al. (2023) did a systematic review of available clinical guidance for MCI and AD identifying critical areas for future advancement. They summarized 15 guidelines for screening, testing and diagnosis, recommendations, and treatment of MCI and AD. They identified the need for the advancement of disease-modifying treatment, which requires constant upgradation of clinical guidelines to align with the latest developments in the field of AD research. The summary of recommendations in advocates no use of biomarkers for diagnosis yet, as the precision of biomarker detection is still low. There is still a wide scope of research to identify the accurate biomarker with precise applicability.

Simultaneously there is increased research in the area of plasma biomarkers, that is, blood-based biomarkers for early detection of AD. Multiple studies are being done to provide evidence for possible clinical applications of plasma biomarkers, as their use can make the screening process much easier and accessible to the general population (Karikari et al., 2020; Pais et al., 2020)

Slow wave sleep (SWS) is another emerging area of research as they are being tested as an intervention for AD. Slow sleep wave disruption is associated with the pathology of AD. Adults with AD pathology have been found to have a low quality of non-rapid eye movement sleep and slow wave sleep in regular sleep cycles. Since these are associated with memory consolidation, memory impairments in AD are associated with low-quality SWS (Lee et al., 2020). Ladenbauer et al. (2017) tested the effects of transcranial direct current stimulation (tDCS) on SWS and sleep-dependent memory consolidation during daytime naps in mild cognitive impairment patients and found positive results. Many studies have found significant positive differences in memory after the application of slow sleep waves. This indicates slow sleep waves can prove to be an effective method for intervention in improving sleep quality, memory consolidation, and associated cognitive functions in AD.

With the increasing applicability of digital technologies, researchers have also employed and tested various digital devices for early detection, intervention, and management of cognitive decline associated with AD. Digital therapeutics is a digital intervention method in which mobile devices, computers, wearable devices, video games, virtual reality, etc., are used (Manchanda et al., 2023). Digital methods such as memory matters and Alzheimer's assistance are used to create an effective rehabilitation environment for improving the cognitive, motor, and sensory impairments associated with AD. Moreover, Bayat et al. (2021) have found that driving with GPS may be an effective and accurate digital biomarker for identifying individuals likely to have preclinical AD. There is an enormous scope for identifying many more such as crucial digital biomarkers from daily used devices which can significantly indicate a cognitive decline in preclinical AD. Later these may also be developed as methods of intervention and markers to assess the progress of the intervention.

Drawing on accumulating evidence, researchers have also identified the gaps in the available literature. In an extensive review of the literature, researchers have found a lack of ethical and culturally diverse data, data based on gender differences and socioeconomic conditions, geographical location, patterns of risk factors, mechanism of operation, and the onset of disease in different areas, which can help in understanding the true nature of Alzheimer's progression in different settings. Future studies in these areas can prove to be of significant help in developing effective treatment and disease management plans (Anstey et al., 2020). It is crucial to focus on and identify the neuropsychological characterization as disease progress varies from patient to patient, depending upon the onset of the disease and their health conditions. It is essential to understand each patient's deficit to give them targeted intervention and provide caregivers with proper support in handling the patient.

Conclusion

There is a lot of research and funding that has gone into developing an appropriate treatment plan for AD. However, there have been a lot of dead ends, with the silver lining being that there is now a greater understanding of the various biomarkers that are a part of the disease. Along with this, there are now multiple scenarios that have been broken down to focus on the amyloid proteins as the earliest signs of a breakthrough. There are many cases where AD has affected not only the person inflicted with it but also the family members. The disease has the capacity to break a person down until their very essence is lost. They are confused and afraid as they are strangers in their own bodies. AD is relentless, and there is very little that one can do. It is a disease that will not be bargained with. There are good days, but there are also horrible ones, but beneath all that, the person is still there. Despite the hardships and the difficulties that come along with it, the loved ones might forget, but they will never stop learning. There is still hope when no one has to see a disease turn their loved ones into shells of their former glory.

References

Alzheimer's Disease Genetics Fact Sheet. (2019). National Institute on Aging. https://www.nia.nih.gov/health/alzheimers-disease-genetics-fact-sheet

Anstey, K. J., Peters, R., Zheng, L., Barnes, D. E., Brayne, C., Brodaty, H., Chalmers, J., Clare, L., Dixon, R. A., Dodge, H., Lautenschlager, N. T., Middleton, L. E., Qiu, C., Rees, G., Shahar, S., & Yaffe, K. (2020). Future directions for dementia risk reduction and prevention research: An international research network on dementia prevention consensus. *Journal of Alzheimer's Disease, 78*(1), 3–12. https://doi.org/10.3233/JAD-200674

Bayat, S., Babulal, G. M., Schindler, S. E., Fagan, A. M., Morris, J. C., Mihailidis, A., & Roe, C. M. (2021). GPS driving: A digital biomarker for preclinical Alzheimer disease. *Alzheimer's Research & Therapy, 13*(1), 115. https://doi.org/10.1186/s13195-021-00852-1

Blay, S. L., Furtado, A., & Peluso, E. T. (2008). Knowledge and beliefs about help-seeking behavior and helpfulness of interventions for Alzheimer's disease. *Aging and Mental Health, 12*(5), 577–586.

Cummings, J. L., Mega, M., Gray, K., Rosenberg-Thompson, S., Carusi, D. A., & Gornbein, J. (1994). The neuropsychiatric inventory: Comprehensive assessment of psychopathology in dementia. *Neurology, 44*(12), 2308–2308.

Dekker, A. D., Strydom, A., Coppus, A. M., Nizetic, D., Vermeiren, Y., Naudé, P. J., Van Dam, D., Potier, M.-C., Fortea, J., & De Deyn, P. P. (2015). Behavioural and psychological symptoms of dementia in Down syndrome: Early indicators of clinical Alzheimer's disease? *Cortex*, *73*, 36–61.

Duke Han, S., Nguyen, C. P., Stricker, N. H., & Nation, D. A. (2017). Detectable neuropsychological differences in early preclinical Alzheimer's Disease: A meta-analysis. *Neuropsychology Review*, *27*(4), 305–325. https://doi.org/10.1007/s11065-017-9345-5

Goate, A. M., Hardy, J. A., & Owen, M. J. (1989). The genetic aetiology of Alzheimer's disease. *International Review of Psychiatry*, *1*(4), 243–248.

Guerreiro, R., & Hardy, J. (2014). Genetics of Alzheimer's disease. *Neurotherapeutics*, *11*, 732–737.

Guzman-Martinez, L., Calfío, C., Farias, G. A., Vilches, C., Prieto, R., & Maccioni, R. B. (2021). New frontiers in the prevention, diagnosis, and treatment of Alzheimer's disease. *Journal of Alzheimer's Disease*, *82*(s1), S51–S63. https://doi.org/10.3233/JAD-201059

Herholz, K., & Ebmeier, K. (2011). Clinical amyloid imaging in Alzheimer's disease. *The Lancet Neurology*, *10*(7), 667–670.

Hippius, H., & Neundörfer, G. (2022). The discovery of Alzheimer's disease. *Dialogues in Clinical Neuroscience*, *5*(1), 101–108.

Hyman, B. T., Phelps, C. H., Beach, T. G., Bigio, E. H., Cairns, N. J., Carrillo, M. C., Dickson, D. W., Duyckaerts, C., Frosch, M. P., & Masliah, E. (2012). National Institute on Aging–Alzheimer's Association guidelines for the neuropathologic assessment of Alzheimer's disease. *Alzheimer's & Dementia*, *8*(1), 1–13.

Jagust, W., Gitcho, A., Sun, F., Kuczynski, B., Mungas, D., & Haan, M. (2006). Brain imaging evidence of preclinical Alzheimer's disease in normal aging. *Annals of Neurology: Official Journal of the American Neurological Association and the Child Neurology Society*, *59*(4), 673–681.

Jones, L., Holmans, P. A., Hamshere, M. L., Harold, D., Moskvina, V., Ivanov, D., Pocklington, A., Abraham, R., Hollingworth, P., & Sims, R. (2010). Genetic evidence implicates the immune system and cholesterol metabolism in the aetiology of Alzheimer's disease. *PloS One*, *5*(11), e13950.

Kalat, J. W. (2015). *Biological psychology*. Cengage Learning. https://books.google.co.in/books?id=EzZBBAAAQBAJ

Karikari, T. K., Pascoal, T. A., Ashton, N. J., Janelidze, S., Benedet, A. L., Rodriguez, J. L., Chamoun, M., Savard, M., Kang, M. S., Therriault, J., Schöll, M., Massarweh, G., Soucy, J.-P., Höglund, K., Brinkmalm, G., Mattsson, N., Palmqvist, S., Gauthier, S., Stomrud, E., … Blennow, K. (2020). Blood phosphorylated tau 181 as a biomarker for Alzheimer's disease: A diagnostic performance and prediction modelling study using data from four prospective cohorts. *The Lancet. Neurology*, *19*(5), 422–433. https://doi.org/10.1016/S1474-4422(20)30071-5

Kozin, S., Barykin, E., Mitkevich, V., & Makarov, A. (2018). Anti-amyloid therapy of Alzheimer's disease: Current state and prospects. *Biochemistry (Moscow)*, *83*, 1057–1067.

Ladenbauer, J., Ladenbauer, J., Külzow, N., de Boor, R., Avramova, E., Grittner, U., & Flöel, A. (2017). Promoting sleep oscillations and their functional coupling by transcranial stimulation enhances memory consolidation in mild cognitive impairment. *The Journal of Neuroscience: The Official Journal of the Society for Neuroscience*, *37*(30), 7111–7124. https://doi.org/10.1523/JNEUROSCI.0260-17.2017

Lai, C. K. (2014). The merits and problems of Neuropsychiatric Inventory as an assessment tool in people with dementia and other neurological disorders. *Clinical Interventions in Aging, 9,* 1051–1061. https://doi.org/10.2147/CIA.S63504

Lee, Y. F., Gerashchenko, D., Timofeev, I., Bacskai, B. J., & Kastanenka, K. V. (2020). Slow wave sleep is a promising intervention target for Alzheimer's disease. *Frontiers in Neuroscience, 14.* https://www.frontiersin.org/articles/10.3389/fnins.2020.00705

Manchanda, N., Aggarwal, A., Setya, S., & Talegaonkar, S. (2023). Digital intervention for the management of Alzheimer's Disease. *Current Alzheimer Research, 19*(14), 909–932.

Mathuranath, P., George, A., Ranjith, N., Justus, S., Kumar, M. S., Menon, R., Sarma, P. S., & Verghese, J. (2012). Incidence of Alzheimer's disease in India: A 10 years follow-up study. *Neurology India, 60*(6), 625.

Nordberg, A. (2008). Amyloid imaging in Alzheimer's disease. *Neuropsychologia, 46*(6), 1636–1641.

Pais, M., Martinez, L., Ribeiro, O., Loureiro, J., Fernandez, R., Valiengo, L., Canineu, P., Stella, F., Talib, L., Radanovic, M., & Forlenza, O. V. (2020). Early diagnosis and treatment of Alzheimer's disease: New definitions and challenges. *Brazilian Journal of Psychiatry, 42,* 431–441. https://doi.org/10.1590/1516-4446-2019-0735

Proctor, C. J., Boche, D., Gray, D. A., & Nicoll, J. A. (2013). Investigating interventions in Alzheimer's disease with computer simulation models. *PloS One, 8*(9), e73631.

Rajasekhar, K., & Govindaraju, T. (2018). Current progress, challenges and future prospects of diagnostic and therapeutic interventions in Alzheimer's disease. *RSC Advances, 8*(42), 23780–23804.

Reisberg, B., Auer, S. R., & Monteiro, I. M. (1997). Behavioral pathology in Alzheimer's disease (BEHAVE-AD) rating scale. *International Psychogeriatrics, 8*(S3), 301–308.

Rosen, W. G., Mohs, R. C., & Davis, K. L. (1984). A new rating scale for Alzheimer's disease. *The American Journal of Psychiatry, 141*(11), 1356–1364.

Sheehan, B. (2012). Assessment scales in dementia. *Therapeutic Advances in Neurological Disorders, 5*(6), 349–358.

Sultzer, D. L., Berisford, M. A., & Gunay, I. (1995). The neurobehavioral rating scale: Reliability in patients with dementia. *Journal of Psychiatric Research, 29*(3), 185–191.

Tahami Monfared, A. A., Phan, N. T. N., Pearson, I., Mauskopf, J., Cho, M., Zhang, Q., & Hampel, H. (2023). A systematic review of clinical practice guidelines for Alzheimer's disease and strategies for future advancements. *Neurology and Therapy.* https://doi.org/10.1007/s40120-023-00504-6

van der Linde, R. M., Dening, T., Matthews, F. E., & Brayne, C. (2014). Grouping of behavioural and psychological symptoms of dementia. *International Journal of Geriatric Psychiatry, 29*(6), 562–568.

What Causes Alzheimer's Disease? | National Institute on Aging. (2019). https://www.nia.nih.gov/health/what-causes-alzheimers-disease

Zillmer, E. A., & Spiers, M. V. (2001). *Principles of neuropsychology.* Wadsworth/Thomson Learning.

Index

Note: *Italic* page numbers refer to figures.

9781032639789